SpringerWienNewYork

Studies in Space Policy
Volume 9

Edited by the European Space Policy Institute
Director: Kai-Uwe Schrogl

Christophe Venet
Blandina Baranes (Eds.)

European Identity through Space

Space Activities and Programmes as a Tool to Reinvigorate the European Identity

In collaboration with
Thomas Ballhausen and
Ulrike Landfester

SpringerWienNewYork

Christophe Venet
Blandina Baranes
European Space Policy Institute, Vienna, Austria

Softcover reprint of the hardcover 1st edition 2013

SpringerWienNewYork is a part of
Springer Science + Business Media
springer.at

Cover Illustrations: © ESA; © CNES/COLLOT Philippe, 2007 (4th row, 2nd picture); © CNES/GRIMAULT Emmanuel, 2011 (4th row, 4th picture)
Source Wikipedia (Kopernikus, Kepler)
Typesetting: Thomson Press (India) Ltd., Chennai
Printing: Strauss GmbH, 69509 Mörlenbach, Germany

Printed on acid-free and chlorine-free bleached paper
SPIN: 80114416

With 32 Figures

CIP-data applied for

Additional material to this book can be downloaded from
http://extras.springer.com
Password: 978-3-7091-0975-5

ISSN 1868-5307
ISBN 978-3-7091-1731-6 SpringerWienNewYork

Preface

The welfare and prosperity of Europe's citizens depends on the internal and external development of the European Union. In a world, where Europe's role and influence are at stake, all efforts have to be made to prevent a downturn. Being a leader in the globalised politics and economy goes hand in hand with an efficient internal setting. This is today's motive for the "Project Europe", it is time to move from the simple protection of peace inside to the maintenance and further development of welfare and prosperity, as far as possible together in a sustainable way with the world around us. In this reasoning, making the European Union work is a matter of crucial importance. It, however, cannot succeed, if it is simply built on abstract rules, procedures and distant authorities. The "Project Europe" can only work, if the citizens join – besides grudgingly accepting institutions – in a European identity. Only this can provide, in the mid- and long-term, a solid foundation for a functioning Union in the inside and vis-à-vis the world.

We are still far away from such a "European identity". And this is why we should appreciate every single option and opportunity for creating and strengthening a European identity, starting with football and ending with song contests. While these examples very much draw on emotions, space provides this and even more. Space provides emotions through astronauts, wonder and astonishment through exploration of distant parts of the universe, excitement through highest technology and acceptance through benefits in numerous policy areas like communications, mobility, resource management and security. Joint efforts such as the European Space Agency (ESA), the European launcher Ariane, the European navigation satellite system Galileo and the European Astronaut Corps signify that space holds an enormous potential to be a showcase for helping create a European identity.

In this setting, the European Space Policy Institute (ESPI) organised a trans-disciplinary conference on "European Identity through Space" on 12–13 November 2009 in Vienna. It was held under the auspices of the Swedish EU Council Presidency. This book contains the further developed presentations of the conference complemented by other contributions and illustrations of the potential, space holds for strengthening the European identity. The readers are encouraged not only to enjoy the enlightening views of representatives from various disciplines but also to use the arguments, which are put forward here to join in the effort for making the "Project Europe" a success.

Kai-Uwe Schrogl
Director
European Space Policy Institute (ESPI)

Table of contents

Introduction

Setting the scene for "European identity through space" as a new field of research and policy-making

Christophe Venet

In his introductory speech to the conference on "European Identity through Space" held at ESPI on 12–13 November 2009, ESA Director General Jean-Jacques Dordain underlined that "the theme of this conference is unusual in space fora, but that also means it is an interesting theme." This sentence summarises the challenges encountered in the process of conceiving the present book, but also the potential usefulness of such an endeavour.

It was first a challenge, as the issue-area of space was long dominated by science and security interests – far away from considerations over identity. Given this relatively narrow definition of space activities, the space community was more or less isolated from other fields of thought for decades. The situation only evolved at the end of the Cold War, when the political relevance of space became more obvious, especially in Europe. From a quantitative point of view first, the number of players engaged in space activities increased drastically. This concerned both public actors (individual states or international institutions) and private/commercial stakeholders. From a qualitative point of view then, space applications started to cross the traditional boundaries of science and security to become increasingly relevant in policy areas such as transport, environment, resource management, development or entertainment. These combined developments clearly increased the political character of space activities. Identity being an inherently political concept, the question of the link between identity and space timidly started to be raised as well. All in all however, the combination of "space and identity" as an object of inquiry and study is almost a virgin territory.

Of course, identity concerns were not absent from the space realm during the Cold War. Space achievements were to prove the superiority of one social system over the other (communism vs. capitalism), they thus had a strong symbolic value, cristallising the two block's identities (East vs. West). The link between space and identity however seemed quite straightforward at the time (space strengthens national pride and fosters a group feeling). The rise of Europe as a new actor in space, and more particularly of the EU since the 1990's, gave a much more complex dimension to the question. The EU itself is still in search of its identity, and this constitutes one of the central challenges – maybe even the most crucial one – of the process of European construction. The complicated nexus of European Space Policy (ESP) actors (ESA, EU, national actors, regions . . .) is perfectly reflecting the multilayered and multifaceted character of the European identity.

Despite the obvious difficulties of dealing with a topic that is both extremely complex and relatively new on research agendas, a book on "European identity through space" also seems to be timely and useful. It could be useful from a research point of view first, as it could set the scene for further investigations and analyses on the relationship between space and identity. As a matter of fact, several authors contributing to this volume are exploring new grounds, laying down first conceptualisations and indicating promising paths of research. This clearly calls for the development of a (small) research field focusing on the relationship between space and identity. From a policy perspective then, it seems timely to launch a reflection on the interplay between space and identity. There is indeed a political convergence between the general development of a European identity and the growing ambitions of the ESP. Both processes are work in progress – there is no clearly defined and perceived European identity yet, and the ESP is still in construction despite evident successes – but in both cases there is also a strong political will to move forward. Given the potential contribution of space to the building of a European identity and the evident importance of identity factors within the ESP, both developments are mutually reinforcing each other.

The mutual interplay of identity and space is precisely the central issue being tackled in this book. The question is investigated by the contributors from two perspectives: how can space contribute to the development of a specific European identity, but also how do the elements of a European identity shape the ESP. The purpose of the book is not to give a definite and unified theoretical framework grasping the concept of European identity, or even of a specific European space identity. It rather intends to illustrate the link between space and identity in Europe from various angles, and to replace the development of the ESP in the broader context of Europe's role and identity on a global scale. Nevertheless, it is necessary to look closer into the concept of identity as a starting point of any reflection on European identity through space. This is the rationale of Blandina Baranes' introductory part "Identity – Attempt of a Definition". This conceptual inquiry into the history, the components, the potential and the limitations of the identity concept provides a framework for the whole volume. In addition, almost all the authors start their contribution with some considerations on identity. It is interesting to note that each author adopts a different approach towards identity, depending on the subject of his/her article.

Interdisciplinarity is indeed another central feature of the present book. In line with the growing political importance of space sketched above, it only seems natural to draw on different scientific disciplines to investigate space-related topics. In the recent years, sociologists, political scientists, economists or anthropologists started to get interested in space activities and to apply the methods and concepts of their research fields to space. Beyond this general trend however, interdisciplinarity was the only plausible approach for the present book, as the concept of identity was mainly studied and analysed by social sciences, meaning outside the traditional boundaries of space sciences.

Specifically, interdisciplinarity is reflected in two different ways in the book. In the first chapter ("Perspectives from Space Studies"), authors with a space-related working background shed light on the link between identity and space *from inside*. They mostly start their reflections from the space field, and then include an identity dimension to their arguments. The diversity here is not related to the difference of research disciplines, but to the different thematic approaches adopted *within* space sciences (international relations and global politics, technology, space industry, space and security, space applications and space exploration). The second chapter ("Perspectives from the Humanities") groups authors with a non-space background who look at the link between space and identity *from outside*. Here, the logic is reversed: the authors start their reflections within their respective research areas (arts and social science, education and communication sciences, anthropology and philosophy), and add a space dimension to their arguments. This crossing of approaches enables to grasp the interrelation between space and identity from both sides: the contribution of the ESP to the emergence of a European identity, and the central role of identity factors in the further development of the ESP.

In addition to the authors' contributions themselves, a number of annexes were included. These are not simple addenda, but should be considered as integral parts of the book, being fully in line with its stated purpose. The first one gathers four short texts written by young Europeans in the framework of an essay competition organised by ESA in 2007. Each participant to the competition had to briefly lay down his personal vision of the value of human spaceflight for European citizens. What is striking when reading these texts is that they equally reflect the rational and utilitarian dimension of human spaceflight (Benjamin Lenoir and Philippa Molyneux) and more inspirational aspects (Daniela Petrova and Regina Peldszus). This highlights the necessity to build European identity on two pillars: a tangible and rational one (common benefits) and an ideational one (common values). The second document in the annex is a timeline of significant events related to the European construction and thus to the strengthening of a specific European identity. It gathers events pertaining to the general political and economic European integration process and milestones specific to the development of a

European space policy. This perfectly shows that both processes are intertwined, and that it is not possible to consider European space policy matters isolated from the overall political evolutions of Europe. The last document finally is the Eurobarometer survey on "Space activities of the European Union" from 2009. Several aspects highlighted by this document are underlined by some of the authors. The most important conclusion that can be drawn from the survey, which is recurrent in the contributions as well, is that European citizens are insufficiently aware of the immense benefits space applications bring to their lives. This in turn, is an important weakness in the perspective of a common European identity in space.

To give a short overview of the content of the book, one needs briefly to turn towards the contributions themselves. As a general and symbolic introduction to the topic of the book, the speech given by ESA Director General Jean-Jacques Dordain at the conference is printed here. While many observers focus on the difficulties for both a European identity and a consistent ESP to emerge, Jean-Jacques Dordain insists on the parallels existing between the ESP and a European identity. Both are challenging tasks, both are children of World War II and both have peace as their ultimate objective. From a purely utilitarian perspective, Europe has no choice but to cooperate in space. This in turn creates common goals which should foster the emergence of a common identity. Besides this rational mechanism however, Europe has the potential to go even further and to develop a "Planet Earth" identity. All in all, this inspiring text is recalling a simple truth: the process of fostering a European identity through space is already taking place, even though this is happening through small steps.

Chapter 1 on "Perspectives from Space Studies", is opened by three authors focusing on the international dimension of the space-identity couple. Michael Gleason first, is investigating the factors bolstering group identity in space. He finds that the ability of space programmes to strengthen national identity is a function of the perceptibility, duration and meaningfulness of the material benefits they bring to their citizens. In the case of the ESP, the two latter elements are given, but the perceptibility of European space activities by European citizens is much too low. Besides adopting a systematic approach to an issue that is often treated in a lousy way, the biggest credit of Michael Gleason's paper is to point out at a central weakness of the ESP, namely its lack of visibility amongst Europeans.

Christophe Venet then, replaces the ESP in the broader context of Europe's international identity. He investigates two questions in this regard: is Europe an international actor in space, and if yes, is Europe a *distinctive* actor in space (i.e., an actor with a distinctive identity). He finds that Europe is a hybrid space power: it is not a radically different actor than other spacefaring nations, but it nevertheless has specific identity features. These have a mediating function on how Europe is defining her interests in space.

Finally, Werner Balogh analyses Europe's activities and influence in the UN Committee on the Peaceful Uses of Outer Space (COPUOS). He highlights the high degree of European involvement in the shaping of international space policies, but also notices that this only takes place in intergovernmental fora: the role of the EU within COPUOS is still unclear and needs to be further defined. Werner Balogh is thus highlighting another key element to take into consideration when assessing the interplay between space and identity, namely the complexity of the European space governance.

Adopting a global perspective as well, Ariane Cornell is investigating whether Europe could be an example for the development of other regional space cooperation endeavours. She concludes that the European model might be hard to reproduce precisely because of the importance of motivational and cultural factors in the success of the ESP. While this seems to indicate that identity does indeed matter in the European successes in space, it also emphasises uniqueness as a central feature of European identity.

Stephan Lingner on his side argues that science and technology are an integral part of the European identity. Even though the European culture was not determined only by it, technology is clearly a cultural practice. Using the example of spaceflight, Stephan Lingner shows that a utilitarian approach is too narrow to justify space exploration, and that cultural elements have to be taken into account. As a whole, he insists on two space-related factors contributing to the European identity: rationality (exemplified by the great European achievements in science) and culture, which might even be seen in Europe as the ultimate goal of space exploration.

Mathias Spude introduces an economic perspective into the debate. He shows indeed how economic integration can lead to the creation of a sense of identity, using the example of Astrium. This integrated European company not only strives at fostering an internal, European corporate identity, but it also contributes to the broader European identity through its achievements in space. This contribution is again showing the importance of the process in the formation of a European identity through space: the slow emergence of a feeling of common identity only takes shape through the daily interaction of thousands of Astrium employees from various European nationalities.

Alan Belward uses the example of space applications (and more specifically of Earth observation (EO) applications) to illustrate two striking features of the emerging European identity on the international scene. Europe is indeed a global benefactor, as illustrated by the use of European EO satellites in the global fight against climate change. At the same time however, Europe is an international competitor: maintaining a competitive industrial base in space is an important goal of the ESP. As a consequence, Europe is trying to increase the competitiveness of its space industry on the global market.

Nina-Louisa Remuss insists on the necessity for Europe to adopt a "formative role" and a "principled identity" in the field of space security. She shows that both participating in an arms race in space and remaining a passive player are no viable options for Europe. She thus calls for the articulation of a specific European identity in the field of space security. Nina-Louisa Remuss clearly suggests that Europe is at a crossroads, and that there is a unique opportunity for her to develop a different approach to international security issues in outer space.

Gerhard Thiele uses human spaceflight as a case study to identify two further elements of identity: values and symbols. Human spaceflight can provide both, first with a series of principles enshrined in the Charta of the European Astronaut Corps, and second by becoming a strong element of European identification.

After trying to identify positive elements and trends highlighting the potential of space to become an important contributor to the European identity, it is also necessary to look at the obstacle to such a dynamic. This is precisely the purpose of Wolfgang Rathgeber's contribution. He warns that the development of the ESP could also bear the risk of endangering the European integration process, and thus the European identity. The lack of progress in certain sensitive issues, such as security policies or the two flagship programmes GMES and Galileo, clearly illustrates this danger. Wolfgang Rathgeber's essay reminds us of two important things. The first one is that there needs to be a conjunction between immaterial factors (political will, common identity) and material structures (a suitable governance architecture) to move forward. The second one is that the development of a European identity in space is not a linear progress, but can be accompanied by setbacks (such as renationalisation tendencies).

Chapter 2, entitled "Perspectives from the Humanities", gives the floor to non-space authors. Giulia Pastorella first, reminds us of the omnipresence of space references in the arts and in the social sciences. Space indeed has always been deeply linked with religion, politics or science, and all these elements are parts of an identity.

Peter Habison then highlights the importance of space education and space-related outreach activities, both to encourage more young people to engage in engineering careers and to foster a common European identity. As such, this contribution highlights the internal dimension of a European identity through space: the ESP has to be attractive not only to the outside world (i.e., on the international scene), but also to the European citizens themselves. A second important idea that could be derived from Peter Habison's contribution is that the process of educating the young generation and the striving for knowledge implied by space education could be considered as part of the European identity.

Moniel Verhoeven and Ignatius van Neerven emphasise the fact that Europe is a fragile "mi-lieu" without a centre, and that its identity is first and foremost related to intangible elements. Both authors argue that due to this specific identity, Europe has a vocation both to explore space and to do this with peace as the highest objective.

Marcus Hornung insists on the participative dimension of identity, meaning the self-perception of this identity by the European citizens. He concludes that Europeans are not aware of the relevance of European space applications to their everyday lives. This lack of attractiveness and transparency is not limited to the space policy field and constitutes a more general challenge in Europe. It reminds that the ESP still needs to gain legitimacy and accountability before being able to contribute significantly to the enhancement of a European identity.

The conclusion, not only of the second chapter, but of the book as a whole, is provided by Jacques Blamont, one of the most distinguished and respected personalities in the European space community. He lays down a bold vision for the future, calling for a radical overhaul of European space institutions and for a focus on the most promising areas, most notably GMES and security applications.

To summarise – and necessarily simplify – what comes out from these various contributions, is not an easy task. A possible way to do so might be to put forward the triangle identity – interests – governance. First of all, most of the authors seem to acknowledge the existence of a specific European identity. Even though it is difficult to define and to grasp, it is there and has the potential to be strengthened (through space). This quest for identity however, should not be detrimental to the pursuit of European interests. The space sector is a positive example in this regard: European Member States took the political decision to cooperate in the framework of the ESP, but they were conscious that the European space industry needed to remain competitive on a global scale. Finally, an adequate governance architecture is indispensable both to preserve European interests and to foster a common European identity. Such an architecture should in particular ensure that European citizens are participating in the policy processes.

These reflections are the result of a long process, which started with the organisation of the conference on "European Identity through Space" at ESPI premises on 12–13 November 2009. This major event took place under the auspices of the Swedish EU Council Presidency, and gathered high-level representatives from the humanities and European space policy experts. ESPI Director Kai-Uwe Schrogl and Ulrike Landfester from the University of St. Gallen served as the conference initiators. Niklas Ström, Counsellor at the Permanent Mission of Sweden, opened the conference. He was followed by two distinguished keynote speakers: ESA Director General Jean-Jacques Dordain and Etelka Barsi-Pataki, President of the Hungarian Chamber of Engineers and former Member of the European Parliament. The first session gathered experts from the space sector: Stephan Lingner, Deputy Director of Europäische Akademie Bad Neuenahr-Ahrweiler; Alan Belward, Head of Unit at the Joint Research Center; Nina-Louisa Remuss, Project Manager at ESPI; and Werner Balogh, from UNOOSA. ESA Astronaut Thomas Reiter rounded up the first conference day with an inspiring and vivid presentation on human space exploration and European identity. The second conference day brought in experts and researchers from outside the space sector: Thomas Ballhausen, Filmarchiv Austria (Literature and Film); Sally Jane Norman, Newcastle University (the Arts); Peter Habison, Director of the Planetarium and Observatories, Vienna (Education and Science Communication); Moniel Verhoeven, intercultural consultant (Anthropology and Philosophy); Joerg Kreisel, CEO of JKIC (Business); and Mathias Spude, Director Communications at EADS Astrium (Industry and Workforce). Wolfgang Rathgeber, ESPI Research Fellow, rounded up the program with a look on critical areas. The conference was concluded by a roundtable with some of the conference speakers and Johannes von Thadden, Vice-President of EADS Astrium; and a résumé of the conference by Jacques Blamont, Adviser of the CNES President.

Blandina Baranes and Thomas Ballhausen, who were instrumental in organising the conference, also summarised its results in a flyer, which can be found in the annex. They identified three distinct themes: European identity and global politics (how space technologies and applications are influenced by, and in turn create a consciousness of Europeanness), European identity and the field of culture (how space, through the help of the humanities, creates other realities that disregard boundaries) and the future of European identity (how space can strengthen European identity). This short document makes a case for "European identity through space" by pointing out the most powerful arguments presented at the conference. As such, it is perfectly in line with the two goals of the conference: shed an academic light on an under-researched field (the link between European identity and space) and provide a practical guide for policy-makers on how to foster a European identity through space.

Given the overall success of the conference, as attested by the quality of the exchanges, the publication of a book appeared as the logical next step. It seemed indeed necessary both to disseminate the results of the reflections nurtured at the conference, and to deepen the analysis. A book on "European Identity through Space" would thus represent the best possible tool to fulfill these two objectives. Very logically, the bulk of the contributing authors were also speakers at the conference. The publication of an extended article offered them the possibility not only to expand on their arguments and theses, but also to integrate the result of the discussions triggered by their presentations at the conference. In addition, the publication of a book was also an opportunity to welcome on board new authors, thus expanding the scope of the analysis. This concerned two U.S. American authors (Michael Gleason and Ariane Cornell), two authors with a social science background (Marcus Hornung and Giulia Pastorella), and one author with a background in international relations (Christophe Venet).

As suggested by the title of this introduction, the present book does not intend to provide definitive answers on the complex interplay of space and the European identity. Rather, it should be seen as the ultimate result of the ESPI conference on "European identity through space", but also as the first milestone towards a more systematic and integrated analysis of two equally fascinating endeavours: the construction of Europe and the exploitation and exploration of outer space. But maybe more importantly, it shall provide policy and decision-makers with a manual on how they can use space for building Europe.

European identity: attempt of a definition

Blandina Baranes

Being on a fact finding mission, the objective is: What is an identity, what is a European identity, whether there is such an identity, is such an identity needed, and if so why?

Promoting such an identity through space is a complicate undertaking. Its pros and cons and the necessity of a "positive" identity in the sense that such an identity inherently contains the idea that we have no single identity and that individuals and groups can belong to many different circles of identities at the same time and that identities can be flexible without being shuttered.

Answering the question of what space could possibly contribute to the formation of an identity and, in particular, a European identity, requires some thought. It is a problematic area of expertise.

Europe itself is a complex concept. On the one hand, it prides itself on its diversity by maintaining its national, historical and geographical borders. On the other, it is increasingly becoming a common economic and political entity. Space expands beyond borders, creating a different state of mind and reaching out for innovation and future undertakings. To consider space as a field for forming and expressing identity requires some courage and visionary thinking. But if one looks closer at this subject it seems not only obvious but striking, that space is one of the most successful identity building tools of the last centuries. Sputnik, the Apollo programme, Yuri Gagarin, the space shuttle flights . . . all of these space undertakings were identity building tools that established national pride and prestige. In Europe similar space undertakings may, to a certain extent, serve this identity building processes, e.g. the Ariane rocket, the European astronaut corps, the scientific laboratory for the ISS, Columbus, the ATV and the scientific achievements we are expecting from the Herschel and Planck satellites – all these space undertakings are purely European and could be part of the creation of a European spirit.

Starting with astronomy and the related scientific areas of the Renaissance and the Age of Enlightenment, it is arguable whether scientists such as Leonardo da Vinci (in his role as a scientist), Nicolaus Copernicus, Galileo Galilei, Johannes Kepler and Isaac Newton up to Albert Einstein could be regarded as pure Europeans. Their scientific achievements have had their influence in the historical, academic and societal sphere of Europe far beyond borders. Astronomy has always been regarded as an academic area of openness, broadening horizons and getting some vague ideas of the universe we are living in. And so does space science of today. Europe in its capacity as the world leader in space sciences and the peaceful use of space is playing a very important role to continue this tradition of identification beyond borders. It shows the uniqueness within European borders versus the nationalism of each European nation.

"... Und in der Tat ist der Weltraum ein wichtiger Bereich, in dem Europa eine gemeinsame Identität konsolidieren und seine Einheit verwirklichen kann[1].... Europa muß seine eigenen Augen und Ohren im Weltraum haben... zur wahreng seiner politischen und auswärtigen Interessen."[2]

This quotation stems from the final report of European institutes on the future of space policy of Europe.[3] This report gives an historical glimpse of the status of space policy in the 1980's. It shows that at a time when Europe was divided between East and West, when the Soviet Union still existed and the idea of a reunified Europe was still a vision, space actually was one of the main inspirations and driving forces to change the whole system. The main message at that time was that Europe can remain a global player in political, technological and academic terms, as it will have a common policy, and in particular a common space policy. ESA as a multinational organisation has become a role model for Europe with respect to contributing to a shared European spirit and a unified European future.

Legitimacy. The outcome and main-findings of the conference "European Identity through Space" provides a solid basis for investigating this rather philosophical, political and technological area of expertise. It not only emphasised for the first time that there is a clear connection between identity, technological achievements and innovation but it also supported further investigation of how a European identity could be formed and which role space could play in this context. Space technologies and space applications were clearly outlined as key elements for a European consciousness. Included within these key elements are: GMES, Galileo and Ariane. Integrating the humanities into this discourse shows that Europe in its diversity can be unique. Assessing the recent past and the near future shows how borders can be transgressed – first in peoples minds and then in political and geographical terms.

Interdisciplinarity. Learning about identity is essentially multi-disciplinary. The perception of space by the public is certainly an interdisciplinary issue as well. It is simply astonishing how space finds its expression in philosophy, the arts, literature, film, education and science communication. The reflection of space by the humanities is remarkable as much as the cognition how European life, politics, economics and education can benefit from space achievements. An extensive view on this discourse as to how European identity could be shaped by space must be interdisciplinary, because only through interdisciplinarity are we able to gain an overall impression of how these two areas are linked with each other. It is a basis for a noticeable discussion with the expectation of obtaining new visions, new facts and new insights.

Processes. From a historical and political perspective, Europe has undergone an enormous change in the last sixty years. This change is highly complex and leaves Europeans with some confusion about their collective identity. What could possibly be the guiding principles of Europeaness? Among others, space efforts are concrete examples of creating a European spirit. Whether through a corporate identity in space, the presentation of the European astronaut corps, or through European cross border cultural undertakings connected to space – space is a role model for finding solid elements as to how we can define and regard ourselves as Europeans.

Influence. Europeans know little about space sciences and space efforts but are aware of the importance and benefits of space for European society. Knowledge about space and space endeavors is requested by the public. Communicating science, communicating space will be the challenge of the forthcoming years. With an interdisciplinary approach and the incorporation of other disciplines we are reaching out to the broader public. Our behavior is influenced by knowledge and understanding. We should ensure that space will not be regarded as something very uncertain or "out there" but should be seen as something which is totally embedded and part of our daily life. Knowledge is the only way to obtain the support of opinion leaders and the broad public. With such support, further research and the financing of science will be not only accepted by the taxpayer but also justified.

The importance of a public understanding of the need for a democratised space study is just as important: given the powerful tools and means that space science withholds, no transparency, supervision and just allocations of means would be enabled or implemented without a strong, vast and ongoing public demand. In other words: a sense of ownership of space-studies by the public is necessary to form an identity driven from science research, and that very identity can – and should – make sure that indeed the space studies benefit the public and stand at its interests.

Defining identity: To achieve a common European policy and a common European future it is of the highest importance to create or to aim for a common European identity.

Space technology, space policy and space achievement throughout modern European history could be an inspiring model for decision makers and for politicians to define Europeaness. Cooperation in space and within space sciences has always crossed borders in people's minds and in their thinking. By thinking about these aspects we are thinking about our lives and we start questioning what we call "culture" and what we call "identity".

By definition "culture" is "the way of life of a group of people" – how people live their lives and how they behave; while "identity" refers to a self perception – *how we think about ourselves, how we think about other people around us and what we think others think of us.*[4] "Identity" simply means "knowing who you are". To analyse the concept of identity we need to employ different disciplines such as philosophy, sociology, history and political science. It is our aim to define a "cultural identity" and in particular a "European identity".

Trying to find answers to such questions as: "What are we?", "What makes us human?" or "What makes me who I am?" leads us to a definition of "identity – something that can be identified".[5]

The philosopher would find the definition of identity starting with the definition of Descartes: an individual with its individual identity is merely "A thing that thinks. ... a thing that doubts, understands, affirms, denies, is willing, is unwilling, and also imagines and has sensory perceptions."[6] The sociologist Richard Jenkins defines identity as "...our understanding who we are and who other people are, and reciprocally, other people's understanding of themselves and of others (which includes us). Social identity is therefore, no more essential than meaning; it too is the product of agreement and disagreement, it too is negotiable... . Whether in the abstract or the concrete, with reference to ourselves or to others, in personal depth or during superficial casual chat, with reference to individuality, nationality, social class, gender or age... . It seems that we cannot do without some concepts with which to think about social identity, with which to query and confirm who we are and who others are...Without frameworks for delineating social identity and identities, I would be the same as you and neither of us could relate to the other meaningfully or consistently. Without social identity, there is, in fact, no society."[7]

Different types of identity: "Identity is an active construction".[8] We can distinguish between an individual identity, a social identity and a cultural identity; belonging to a "symbolic community" or an "imagined community" are keywords in the vocabulary used in this discourse.

Postmodernists relate to identity as becoming "increasingly freer, ambiguous and plural in the postmodern age... . The master identities of class, age, gender and ethnicity are decomposing: they are falling away to be replaced by new identities based on the whole range of sources, including consumerism, the body and sexuality".[9]

Identities are shifting from individual and group identities to more open identities driven by the force of consumption and no longer by production as it has been during the last two centuries.[10] But the argumentation reaches out for a new definition of a "consumer identity" and it is not only the materialistic way of consumption which is meant but also the identities based on "cultural capital". Education, progress, innovation can be regarded as cultural capital at which young Europeans are aiming. In this sense it could be argued that space could fulfil exactly these needs and could be part of creating one of these new identities branching out from old models – identities that are determined by socialisation, education, interest and profession – remaining in the European culture as a "macro" pattern, but creating "European identity" as a "micro pattern" beyond societal borders, ethnic and religious roots.[11]

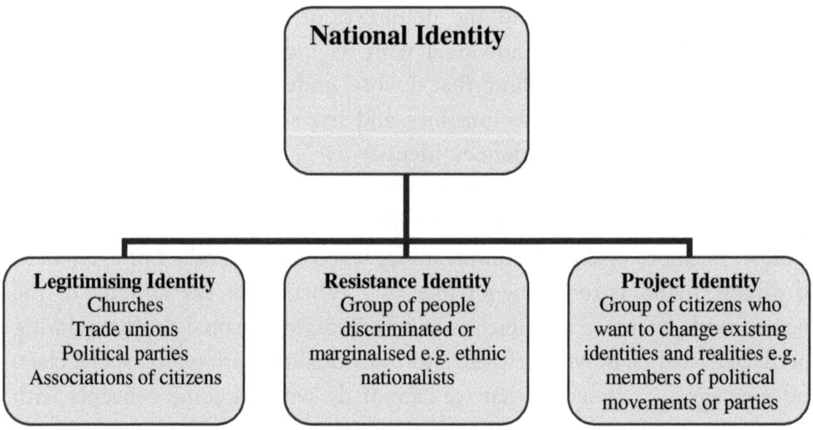

Identity building: Being a scientist, an astronaut or a politician implies a social role but it could also be an identity if the profession is an important issue of self-definition. It demonstrates that an identity not necessarily can be defined only by a national identity but can be perceived as how someone defines himself. Language, job, education, religion etc. can as such influence self-definition. Identity is not something that is unalterable, it is a subject of evolution, a process of amendment through changes in social roles and it is even something which could be influenced and shaped by political strategies.[12]

According to Cristina Chimissio there are four elements in creation of a European identity:

- History: collective memory and historical development as a construction element for identities
- Geography: to identify national identity it is essential to define the territories
- Religion: strong identification model, especially for minorities within Europe
- Language: official language(s) of a country are a fundamental indicator (e.g., English being an official language in many countries versus Polish being the single official language of just one country); also an accent signifies being part of a certain segment of society or the belonging to a certain geographical region – both are very strong identity forming microcosms

Let us take the opportunity of thinking of another element which could to a certain extent contribute to the formation of a European identity:

- Space

The European Astronaut Corps is a case in point. Over the years the wording identifying the astronauts has changed. Having been addressed by nationality during the past years, astronauts of today are addressed as European astronauts with his or her specific nationality. This points out a clear message – we are Europeans with our European Astronaut Corps. Space education, space technology and the creation of perspectives in space cannot be envisaged within national borders. Education and in particular space education can be a very useful tool for implementing European identity. Teaching history has shown the direct impact on identity building processes. Teaching space sciences or space policy could serve the same interest for future visions of Europe. It should be the aim of educational systems in Europe to create Europeans.

Culture and identity: European identity on first sight would be defined as a "cultural identity" while from other perspectives it could be defined as class or national identity. The nation building process in Europe was a remarkable example how new identities have been created throughout European history.

From the anthropological point of view modes of thinking, feeling, behaving, but also values, customs, traditions and norms are not only part of a certain culture but also gives a framework of identification of individuals. The conception of "culture" is changing, is influenced and certainly dependent on external factors. Our views on religion, philosophy and visions of human destiny are changing.[13]

European identity – does it exist? When considering its diversity, Europe might leave the impression that a European identity does not exist, not visible, not perceptible and definitely unreachable. One might answer the question whether a European Identity exists with a clear negative reply. Europe as a nation –

unthinkable since a nation is determined by a common history – a shared past, by language by traditions and geographically.

But referring to the timetable (see: Appendix 2) illustrates the development of Europe from a continent of borders to a continent without frontiers. Barriers have fallen over the past years – economically, strategically and in people's minds. It is the young generation that profits from this freedom from barriers. The educational programmes of Comenius, Erasmus, Leonardo da Vinci and Grundvig which provoked a change in mobility and in narrow-mindedness led to a Europeaness and a completely different consciousness of being European. In 2006/2007, approximately 160,000 students participated in these programmes and to quote a student "... that experience made a whole new person of me and that I would never look at the world and Europe, my home, as I did before." It shows that the idea of being European is being implemented by these educational life-long learning programmes. European identity is a state of mind and finds its induction in these educational programmes.

Why at all a European identity?

In regard to space policy European identity is a value that should not be underestimated. The key factor in Europe is cooperation. Europe can move forward through cooperation and remains a global player. In its diversity Europe's chance is to make use of the advantage of cooperating with different players. One hindrance in Europe might be that especially in space programmes there are different priorities in different countries. But still space policy is a role model in terms of cooperation, acting together and creating a European spirit.

But the diversity is also one of Europe's most substantial advantages. Diversity can be the ongoing source that generates creativity, sensitivity, vision. Diversity can create the infrastructure for social mobility, and promote openness, accountability and transparency.

These two tables of Eurobarometer surveys conducted in 2006 and 2010 show that self-identification and belonging to a supranational polity like the European Union is still a very fragile matter.[14] People would feel rather identification with their home village and their home country than with the European Union.[15]

"I feel British ... except when I am around people from other continents ... then I feel very European ... ." This blog entry at a discussion platform where the question "Do you consider yourself British or European?" puts the attitude of European citizens of various nationalities in a nutshell. But the wish to make the European Union more an active actor for all European countries in various policy

In the near future do you see yourself as...?
06/2010

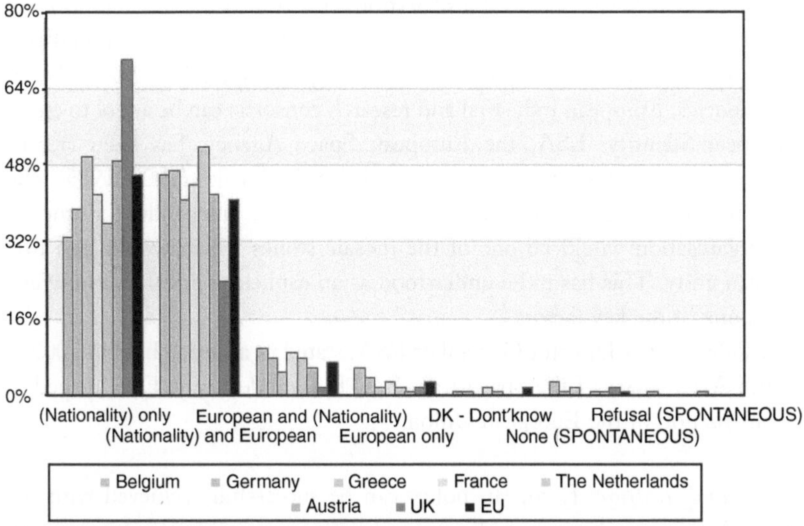

Do you ever think of yourself as not only (nationality), but also European?
Does this happen often, sometimes or never?
09/2006

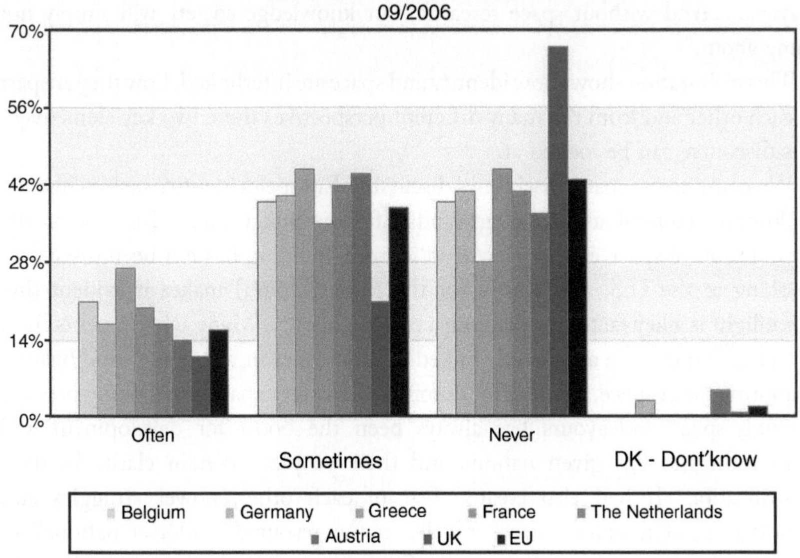

areas such as energy, defense and in particular in education is prevailing. What creates a "passionate European"? This question brings the focus on how this idea of the "European citizen" could be realised one day. It is the experience of friendship, communication and common goals.[16] Common goals conducted through European Agencies, European industrial and research consortia can be a tool to create a European identity. ESA, the European Space Agency, has been created according to the philosophical approach of Edoardo Amaldi (1908–1989) who for the first time argued that the establishment of an independent European space organisation would be one of the mosaic stones which would lead to a European unity. This has to be understood as an explicit political aim in which space is one of the key factors.[17]

Roy Gibson, first Director General of ESA, stated in a speech held in 2007 at the 50th Anniversary of the Space Age. "… In a modest way, ESA helped in the construction of the European Union."

Synthesis and structure: European policy can be successfully achieved with the common thread of a European identity and a European spirit.

European space policy can be successful if there is a clear statement from the political bodies as expressed by José Manuel Durão Barroso President of the European Commission at the conference entitled "The Ambitions of Europe in Space Conference on European Space Policy" held in Brussels, 15 October 2009.[18] "Space is also essential to project the image of the EU as a world player…. And without space research, our knowledge society will simply not come about."

This elaboration shows how identity and space are interlinked, how they are part of each other and from the many different perspectives these two key elements of this discourse can be looked at.

We come across many disciplines discussing European identity such as history, philosophy cultural anthropology and different policy areas. But so far the connotation of space as a key factor in identity building has not been discussed. Looking at the U.S. and Russia (or the Soviet Union) makes it evident that spaceflight is a key factor for defining a certain identity. Many achievements from spaceflight and space are directly linked to milestones in technology and history. Exploring outer space, fighting for dominance in outer space and gaining prestige through space endeavours has always been the boost for development and innovation and has given nations and their people a certain clarity in their identification. It had also created fear of each other, power struggles and exploitations of human resources as well as nature resources, endorsed nationalism and suspicion, and enhanced processes of armament. Europe, thus, can and should carefully learn from both the positive and the negative roles that space had played in the past in the creation and the promotion of publics' identities. As result space enhances the European identity, serving as a role model for many other policy areas while also creating new realities.

Structuring this book was a difficult undertaking as grouping the contributions and classifying them was a big challenge. Should the classification be from a philosophical point of view, from an "inside/outside" view, from a technology versus a humanities view? Finding the answer was difficult but it led to a striking cognition: space has no borders and dealing with space makes a clear concept of classification nearly impossible. There are no borders and that's the fascination of space – it leaves space! Space for thinking for visionaries, for the facts achieved by the natural sciences and for ideas created by the humanities. This has been exactly the approach of this conference and it was amazing to observe scholars from different disciplines find a common language within this context. An enriching dialogue was the result and will be presented in this volume of the "Studies in Space Policy".

The European future and European heritage is strongly shaped by knowledge and technology. How we perceive technology is a question of how technology is communicated to the public and how we sense technology. Symbols are strong components in this communication process. Space offers these needed symbols. The Ariane rocket, Galileo navigation system and in the near future GMES are clear defined expressions of European technological knowledge and accomplishments. Looking at the 20th century, European technological advancements were in general related to war and to military requirements. The change of paradigm took place during the post war era. Not military but civilian use of technology and in particular of space can be regarded as a "gift" of this century. It is the task of the Europeans to remain in this tradition and to foster research and development as in contrast to ignorance and lack of interest which would lead us to dependence and backwardness. Europe as a global player cannot afford such a status in the world. Space industry such as Astrium, contributes to a corporate identity in Europe.[19] Space exploration and space travel, e.g. Virgin Galactic and Space X programmes, are spirited and inspiring European endeavours where a cultural development is certainly evident. Through the Lisbon Treaty scientific and technical progress in Europe is advanced – space policy is explicitly mentioned as one of the hubs for this envisaged progress.

"Spaceflight is a cultural enrichment"[20] and "space gives an unlimited inspiration to art".[21] These two statements encompass the whole range of the exciting field of technological and cultural cross over. The second chapter of the book looks into the details of the role space plays in this context. European literature, film, art and, one would argue even philosophy, has been inspired by space. Space is a catalyst for future generations in terms of science, education and communication.

President Obama addressed the U.S. Congress in his speech at the beginning of 2011 and mentioned the "Sputnik moment": how many words are needed to describe the immediate connotation by mentioning "Sputnik". Is it a wake-up call, is it a plan "win the future", is it the expression of fear that other could overrule, is it a strike against America's national pride? Is it a catalyst for future science and technology developments? Certainly using the term "Sputnik" has a very strong meaning and a very clear message. It illustrates that expressions taken from space have an enormous impact. Space stimulates and expands our imagination. For mastering the future, in particular our European future, we need these visions and a wide horizon we need for mastering the future. It is our curiosity which leads us to reach out for new horizons.

Space is not empty. Space is filled with imagination, vision, ideas and ideologies. Since space shifted from being a political instrument of governments to an instrument being used by a majority of individuals on this planet we have to rethink and to redefine space. Incorporating the humanities in this discourse will create much broader scope to create awareness of the benefits of space.

Our imagination of outer space has been significantly influenced by the creativity of writers, artists and musicians. Will spaceflight, exploration and space sciences be part of a European or even global future? It will be the task, not only of politicians and scientists, but also of artists to create willingness to continue these undertakings. Space should not turn out to be a myth of the space age but it should become again the driving force for science technology and economy. As Kennedy said in his famous Moon speech on 12 September 1962 "... we do not intend to stay behind, ..., we shall make up and move ahead"[22] Europe has to prevail in its position as a leading launcher, also as a leading cultural launcher. Science fiction with its pre-visional thinking prepared the ground for the development of innovative technologies. Europe is more likely than other nations to show how soft societal values are linked up with hard technology values. Philosophy, the arts, educational systems and communications are a constant and provide a wide platform of exchange.

Arthur Woods states: "The future of space activities, the future of humanity and perhaps even the future of all life on Earth is in need of skilled communicators possessing the knowledge and understanding of the scientist combined with the intuition and sensitivity of the artist."[23]

Spatial culture in different expression such as visual arts and sculptures (the "Cosmic Dancer"), the cultural utilisation of the ISS, music and literature are "transformers" of dreams and visions. The humanities are the voice of desire and as a result, the realisation of reaching out to new horizons bringing us the knowledge of the Universe. Barriers are stopping us from advancement, discoveries are encouraging us. Space through a humanistic loom becomes a reality.

The main findings of this survey are diverse. To put it in a nutshell, according to the statement of former Commission Vice-President Günter Verheugen the main message is: "Without the European Space Policy, Europe could become irrelevant. With this Resolution on the European Space Policy, we intend to live up to Europe's global leadership aspirations in important industrial and research areas, which will provide growth and jobs for the future. Moreover, space has always been a source of inspiration which helps people to think outside and beyond the limits and to innovate. Today's proposal marks a milestone, to ensure that Europe does not miss out on the important opportunities that space technology offers."[24]

European citizens are aware of the importance of space efforts. Space as a source of knowledge, of investment, of innovation and of prestige is undoubtedly perceived by the European public. The benefits from space for environmental protection and disaster management were considered as very important. But there is a huge lack of information concerning concrete knowledge about space sciences, space programmes, satellite services and space exploration. The information that reaches the European citizens is rather diffused without putting it in the context of innovation, economic stimulus and future development in Europe. That the EU should do more in the field of space exploration was agreed by a slim majority while 30% expressed a negative attitude towards space exploration. In terms of budget planning the majority agreed that budgetary resources for the upcoming years should be maintained and not decreased despite the financial crisis. One of the stunning results of this survey was the fact that a large number of respondents hardly knew about the European Earth Observation satellites. In some countries a very high percentage of interviewed persons had not even heard of such European satellites, such as in the UK 73%. Many respondents were on the other hand familiar with space applications for environmental monitoring, communications, positioning systems and security. Concerning the question whether the EU has to increase activities in regards to space exploration the thoughts were divided half against and half in favour of more involvement in space exploration. The main message from this survey is that Europe and space are two perceptions which are invisible and totally abstract for the public. In order to make people realise the importance of both, communication must be improved. Historically Europe and space are interlinked by two milestones in history: 1) the crucial year of 1957 for Europe: 2) the Treaties of Rome were signed and it was the year when Sputnik was sent to space by the Soviet Union. Both historic moments were not only perceived by the public but also remained in people's memory. These were definitely two moments of identity building.

One of the milestones of European Space achievements was the installation of the Columbus laboratory at the ISS in 2008. This ambitious undertaking was accompanied by an essay competition of ESA, where young students from ESA Member States were invited to express their point of view on the value human spaceflight has for European citizens. This competition was conducted in 2007.

Since this theme touches the topic how space is on the one hand reflected by young people, and on the other hand which visions are behind such projects, this publication contains in its annex the winners contributions in this essay competition. Columbus being a purely European endeavour could and should be one of the "identity building" measures which ESA has to emphasise towards the European public in order to show its part in space.

"A role-model for aspiration . . . that will lift [them] to new heights . . . the value of human spaceflight until now is unquestionable and with the Columbus' laboratory it will grow increasingly valuable . . . "[25] The contribution of Daniela Petrova contains in a very lively and poetic manner, how much space achievements contribute to our daily life and how much people are benefitting from these achievements in regard to education, medicine, disaster management, environmental awareness, energy generation etc. At the end she brings it to the point of how Christopher Columbus by reaching America changed the world in terms of crossing borders in minds and geography. "Like the tamers of the land and sea we tame the value in air and space we see."

Political and economic aspects of human spaceflight have been emphasised. Europe can compete with the United States and China just as long as it operates as a strong and homogenous entity. Being competitive by being together, being economically strong, being open for innovations and led by visions, that is the future and the importance of human spaceflight versus unmanned missions as it is seen today from the viewpoint of this young student. The essence of this essay is that it is of utmost importance for young people to know and to rely on politicians and their competence to ensure and to secure education and research, where human spaceflight is one of the most ambitious ongoing projects. As seen through the minds of young people it is exactly what ensures higher standards of living for European citizens and contributes to the economic benefit and security of Europe. Leadership in Europe in terms of education, research and in particular in space development and space sciences is claimed by the young generation.

Another contribution is shaped around the "state of mind" by emphasising a sense of openness. Curiosity, creative potential, problem–solving, the ability to imagine and to improvise, anticipation and inspiration makes us move on. Spaceflight offers many possibilities to recognise these abilities and through them we are able to improve and to re-evaluate the status quo. The European Union and the joint space programme are achievements where a vision became true through openness and courage to cross boundaries.

Another focus is on the political and economical aspects of human spaceflight. Europe could gain again sovereignty through a more knowledge-based economy where technical and scientific expertise is one of the key factors. Inspiration through space and human spaceflight is certainly an encouragement for young people to choose studies in technology and science which is definitely needed in the near future in Europe. She also makes in her essay a clear statement in favor of human spaceflight vis-á-vis robot explorers.

The essence of this essay competition is certainly the expression of the wish of young people to have guaranteed their possibilities of development in terms of education and research in the near future. Columbus stands for the historic person of Christopher Columbus, reaching out for new horizons, but Columbus stands also for new technical accomplishments that will lead to new scientific achievements. This historical perspective combined with future visions of science and technology in particular in space sciences and human spaceflight contains the main message of this contest. And it is a European undertaking – young Europeans were addressed for showing their deep interest and their understanding how the future in Europe could be shaped by space. It is a case for clear European leadership, a leadership through capability, a leadership through space.

---

[1] Europas Zukunft im Weltraum: Ein gemeinsamer Bericht europäischer Institute. Bonn: Europa Union Verlag, 1988: 1, 168

[2] *Ibid.* „... in fact space is an important domain, in which Europe can consolidate its common identity and realize its unity.... Europe must have its own eyes and ears in space... in order to maintain its common political and international interests..."

[3] Forschungsinstitut der Deutschen Gesellschaft für Auswärtige Politik (Bonn), Institute Français des Relationes Internationales (Paris), Istituto Affari Internazionali (Rom), Nederlands Instituut vor Internationale Betrekkingen „Clingendael" (Den Haag), Royal Institute of International Affairs (London)

[4] Kidd, Warren. Culture and Identity. New York: Palgrave Macmillan 2002: 7

[5] Chimissio, Cristina (ed.) Exploring European Identities. Oxfordshire: The Open University, 2005: 6

[6] Descartes, René. Meditations on First Philosophy, with Selections from the Objections and Replies, (1641) (translated by J. Cottingham) Cambridge: Cambridge University Press, 1986

[7] Kidd, Warren. Culture and Identity. Hampshire: Palgrave Macmillan, 2002: 25 (see also: Jenkins, Richard. Social Identity. London: Routledge, 1996)

[8] *Ibid.*

[9] *Ibid.* 27

[10] *Ibid.* 94

[11] Based on: Castells Manuel. The Power of Identity. Oxford: Blackwell, 1997.

[12] Chimissio, Cristina (ed.) Exploring European Identities. Oxfordshire: The Open University 2005: 34

[13] Dick, Steven J. Space, time and aliens. The role of imagination in Outer Space. Abstract. *http://www.geschkult.fu-berlin.de/e/astrofuturismus/Bielefeld/abstracts.html*

[14] European Commission: Eurobarometer Survey 2010/06 and 2006/09

[15] Melchior, Josef. The Transforum Essay. The Present and Future of European Identity: An Experiment in Para-Nationalism? *http://www.acfny.org/transforum/transforum-1/the-present-and-future-of-european-identity-an-experiment-in-para-nationalism/*

[16] Grundy S. and L. Jamieson, 2007: 677

[17] Dick, Steven J. and Launius Roger D. (eds.) Societal impact of spaceflight. Washington: NASA History Division 2007: 593.

[18] *http://europa.eu/rapid/pressReleasesAction.do?reference=SPEECH/09/476&format=HTML&aged=0&language=EN&guiLanguage=fr*

[19] Spude, Mathias. Does the European integrated industry contribute to a new European identity? (Chapter 1.6)

[20] Thomas Reiter at the conference.

[21] Sally Jane Norman at the conference.

[22] *http://er.jsc.nasa.gov/seh/ricetalk.htm*

[23] Wilson, Stephen. Information arts. Intersections of art, science, and technology. Cambridge, Mass.: MIT Press 2002: 262.

[24] *http://www.esa.int/esaMI/About_ESA/SEM4UU8RR1F_0.html*

[25] Petrova, Daniela. The value of human spaceflight for European citizens. www.esa.int/esaHS/SEMR5353R8F_education_0.html (winner of the ESA Columbus essay contest)

A European identity through space – what have we achieved?

Jean-Jacques Dordain

Thank you for inviting me to this Conference organised by ESPI here in Vienna. ESA and Austria are founding members of ESPI, and I think it was a very good decision to create ESPI, whose importance is growing. The theme of this conference is unusual in space fora, but that also means it is an interesting theme. In fact, the question of identity is a difficult, and even sensitive, subject. In some countries in Europe there is currently a debate on national identity running. As Director General of ESA, you could say that I need to have 18 nationalities – and the number is growing . . . I'm not an expert on the subject of identity, so I won't try to give you an answer, but I'll tell you about how I think space can bring something to our identity.

First of all, I would like to underline two important events in space, which dramatically changed the world and people's perceptions of identities. The first one took place in 1957. The launch of the first satellite, Sputnik, opened the door to the Space Age. As can still be seen in newspapers from that time, Sputnik was perceived as a demonstration of superiority of Eastern technology. In fact, many countries identified themselves with the Sputnik achievement: the USSR, but also countries of Eastern Europe, creating a common identity among them. Today, 50 years later, that type of identity is out of date.

Another important event took place in 1957: the signing of the Treaty of Rome. This was the basis for the development of what was to become the European Union. In fact, "space" and "Europe" share a lot of common aspects. To start, both are difficult subjects. It is difficult to go to space, and it is difficult to build up Europe. And for both, we have no alternatives. Both are also children of the Second World War. Both have the common objective of peace. The Outer Space Treaty requires that space activities are for peaceful purposes. And the best thing the EU has done is to bring peace to Europe. The last common aspect is that both "space" and "Europe" are invisible. You can't touch space, and you can't see Europe, since they are both rather abstract fields for the public, and so it is difficult to make citizens realise their importance and to build up an identity on these two values.

However, both these historical events stimulated an identity: an Eastern identity and a European. It was in the wake of these two events that ESA was created – in fact, ESRO was created in 1964. So ESA is at the crossroads of space and of Europe. It has the difficulties of space and the difficulties of Europe, so it is twice as difficult! But it is also twice as successful.

Before speaking about achievements of ESA and of Europe in space, I would like to come back to the second historic space event. In July 1969, the first landing of man on the Moon took place. The image given was that of superiority of Western technology. But I am not so sure it has created a Western identity. Because, in fact, it was the start of the end of competition in human space flight. Nothing was the same after that achievement. Thirty years later, nobody really cares anymore that it was the US flag that was placed on the Moon. But the Moon landing meant two real breakthroughs, which were not perceived as such at the time, hidden by the event itself.

Firstly, it was the start of international cooperation in space. There was one flag on the Moon in 1969. In 1975, two flags were represented when Apollo and Soyuz docked for the first time. This was a premonition of the fall of the Wall between East and West. Then, we had four flags in the space station Freedom project. In 1993, these became five with the International Space Station. And I am sure more flags will join, I hope it will be already on the ISS. Indeed I am the one who has been repeating that we should invite more partners to the ISS project.

The second breakthrough was the concept of Planet Earth. For the first time, man could see the Earth as a small golf ball floating in the Solar System. Bill Anders said that "we came all this way to explore the Moon but the most important thing we have discovered is planet Earth". And indeed, our future can only be global. Earth is like a small spacecraft, a finite place, with finite resources. All this shows that what comes out of an event at the time of the event can be totally different from what is seen to have come out of it thirty years later. The notion of identity is not permanent, it is evolving.

I am convinced that space contributes to the development of the European identity. Still, we have a lot to do, and we also have to go further and develop a "Planet Earth" identity or a global identity. This is our future, and it is certainly global. Antoine de Saint Exupery wrote that your task is not to predict the future, but to enable it. We have to work to make our future possible on Planet Earth. There is a space identity, and I want this to help develop the European identity, but I also would like to see that our European identity helps develop a "Planet Earth" identity.

So, what is this space identity? What are its unique features?

First of all, space activities are focused on projects. And we need common projects and common goals to develop an identity. It is then easy to integrate people of different nationalities. When ESA landed Huygens on Titan, a moon of

Saturn, no one in the ESA team thought about which nationality they belonged to, no one in the industrial team thought about which organisation they belonged to. All were working together, as one project. That day you could really see the space identity – men and women crying, because we were the first to discover this world, which is almost an Earth-like system.

I dare say that we have invented an identity in the project teams – and this is not about a French working with a German etc, but really about a new identity. Yes, we speak in English, but it's a very special English . . . . We have symbols of this identity: such as Ariane. By the way, Ariane was used as symbol of Europe in a French election campaign for the European Parliament. The European Astronaut Corps means there is a single corps of astronauts in Europe. So there is an identity of the European astronauts. But when our European astronauts work onboard the Space Station they are first and foremost citizens of the world: no one up there really cares about their nationality or who they work for, they work together as one team with common goals.

The second characteristic is a culture of risk. We shall not be Europe without taking risk. The principle of caution can't apply if you want to make significant progress. In space, you have to take risks. At each launch there are reasons not to launch, but we have to launch. We have failures, but they are part of the progress: we have to manage the risks. Therefore, we use the best talents and expertise to be found: within European industry, within the scientific community – and from 19 different nationalities (including Canada). This is our way of using diversity. I think it is a richness of ours. Space is certainly a very good example of how to manage risks.

The third characteristic is that in space Europe has no alternative but to cooperate. In fact, space activities in Europe are one of the most integrated activities: more than 50% of space activities are on the European level (for research the figure is close to 5%). The new competence on space of the EU under the Lisbon Treaty will certainly increase that level of integration of space activities in Europe. I have heard that Europe does not exist because there is no phone number to call Europe. In the field of space, Europe has its phone numbers. The USA and Russia know which phone number to ring to reach the European partner on ISS (which is to ESTEC by the way). For operations, they call ESOC: the teams there have rescued both a Japanese and an American satellite. If there is a natural disaster: please call ESRIN. This integration is a reality and space is in advance of many activities.

The next characteristic is the well experienced governance in space. Yes, it has evolved and it is still evolving, but it has always led to success. With a much smaller budget in Europe we are in a leading position on a lot of activities in space, meaning our governance cannot be that bad.

The last characteristic is that space inspires the young generations. And we shall make Europe with the young generations. Space and Europe are two concepts that will grow, in particular thanks to the young. So we have to attract the young, and space is doing just that.

This is the way our space identity can build up our European identity.

So, how can the European identity build up a "Planet Earth" identity? Are there any such characteristics of Europe in space?

Firstly, the European space sector has been developed as a civilian sector, which is different from the other space powers: USA, Russia, China. Japan is the only one more like Europe. This means that priority has been given to science and to services to the citizens, which is pretty unique in the world. That's why, in spite of lower budgets, we are on the leading edge of space science. We have recently launched Herschel and Planck: there is no equivalent in the world. The data now coming from these spacecraft is unique (we still have to wait a little for the data to be turned into science though). Europe is also on the leading edge on Earth Observation activities. We have today 20 Earth Observation satellites in development in ESA. SMOS was launched only last week, to study the water cycle, with unprecedented accuracy. And in the telecom domain, we have the most successful operators in the world. SES global is one example. For launch services, we have Arianespace. Even though we are the smallest space power of the world, we are among the most successful in space.

The second characteristic of Europe is the well experienced complexity. I have said before that it is hopeless to try and make Europe simple. So let's use the complexity and diversity. But in fact, the organisation of space activities in the US is not so simple either. And to understand how the Chinese space sector is organised would be very interesting... We have in Europe 50 years of experience, with the regions, nations, the intergovernmental and communitarian levels... In the end, it works and this diversity is one of our strengths.

Lastly, Europe has a long history of relationships with all parts of the world, which no one else has. Especially with the parts of the world which are not space powers, but which still need to use space. This is one of the strengths of Europe that we have such relations established with South America, Africa, the Far East etc. We are possibly the only space power cooperating with everyone, including with the non space-faring parts of the world.

So there are unique characteristics in Europe which can open a way for Europe to develop a "Planet Earth" identity. True, there is much more to be done than we have done, so I would like to point out the three most important challenges, in the short, medium and long term:

In the short term, space must contribute to the economy and its recovery. This is a global challenge. It involves both competition and cooperation. Take for example Galileo, I think it should not be seen as a competitor to GPS – rather, the GNSS are global systems to help navigate on Planet Earth. Space and Energy is becoming more and more important. In the short term, the EU is looking to the power grid, and here space can do a lot. We have energy coming from the winds to the North, from the Sun to the South, from the gas and oil to the East... Space can certainly bring a lot of information for regulating this power grid.

In the medium term, we have Climate change. When I say medium term it does not mean that we should not act now – this is a pressing issue. And it's a global issue. It is even more complex than space, since here we have a whole chain of actions, starting by collecting data, modelling, predicting, acting, monitoring... But what is striking is that space is on both ends of the chain: able both to collect data and to monitor the consequence of the actions. At ESA we have a fantastic programme to collect data: the Earth Explorers: one satellite launched per year. Also our missions out in the Solar system can help: there was a dramatic climate change on Mars, and this can help us understand also Planet Earth. There are also the services, the monitoring part, such as under GMES, where the European Commission is leading with the support of ESA. On these services, Europe is on the leading edge in the world.

Long term, we have Exploration. The recent conference on human space exploration in Prague meant to start the debate. A question asked was: What is exploration? I think a good definition is: extending the knowledge and the actions of humans beyond Planet Earth. The future of the Earth cannot be seen in isolation, it is part of a wider system. In fact, Exploration may be the only global endeavour we could attempt without acting in response to a crisis, such as Climate change. An international Exploration initiative will start with the success of ISS, and we still have to make sure that ISS becomes a success also of utilisation. It would be very difficult to ask Member States to invest in exploration if they saw the ISS investment as useless. Therefore I am glad that the US has taken a decision towards extending exploitation to 2020. And exploration is both robotic and human. In ESA, we put particular effort on the robotic exploration of Mars. Therefore, I am particularly glad that ExoMars is not anymore a one-off mission, but part of a long-term cooperation with NASA.

To end, I must point out that I am an engineer, not a philosopher – and certainly not an artist! – but I have here given you my personal perception of the debate you are organising. I want to thank you for this opportunity. This

Jean-Jacques Dordain, Director General of ESA (left) during the opening of the conference together with other keynote speakers from left: Etelka Barsi-Pataky (MEP), Harald Posch (Head of FFG's Aeronautics and Space Agency), and Jaques Blamont (CNES)

Conference and these discussions are certainly a way of associating the public to space and Europe, which I am sure is very important for driving the future.

CHAPTER 1

PERSPECTIVES FROM SPACE STUDIES

International perspective

1.1 Space, pride, and identity*

Michael Gleason

Many have claimed that there is a relationship between space and group identity. Walter McDougall in his classic tome *The Heavens and The Earth: A Political History of the Space Age* and James Oberg in *Space Power Theory*,[1] draw an analogy between Frederick Jackson Turner's 1893 thesis on the importance of the frontier to American identity and the role that space now plays as a new frontier upon which American identity may be continuously forged. Oberg also asserts that nations use space activities for inspiration and affirmation of national consciousness, expressions of national character, binding together, internal order, and as a potent symbol with a futuristic aura. McDougall goes so far as to say that space programmes "define mankind in the first place".[2] In *Pale Blue Dot*, the popular astronomer, Carl Sagan, surmised that space has the capacity to unify all mankind.[3] More recently, the 2007 European Space Policy states that space "can contribute to European cohesion and identity"[4] and the 2006 Chinese White Paper "China's Space Activities in 2006" refers to China's space programme as "a cohesive force for the unity of the Chinese people".[5] Clearly, space policy experts and scholars assume that a casual relationship exists between space programmes and group identity. Nevertheless scholarly justification for such assertions is sparse.

This study attempts to begin filling that gap by first laying out the factors that bolster group identity. It is found that the most effective way for leaders to strengthen group identity is to provide meaningful, perceptible, and enduring material benefits to their group members that are also visible to those outside the group. The provision of such benefits creates an individual's sense of pride in the group, enhances the group's prestige among those outside the group, and in turn, strengthens group identity.[6] In this regard, space programmes do have the potential to instill pride and prestige for the group and thereby bolster group identity. But it is argued that the effect space programmes have on identity will vary with various types of space activities. For example, collective European space programmes prioritise their space activities differently than the Chinese.

*The views expressed in this article are those of the author and do not reflect the official policy or position of the United States Air Force, Department of Defense, or the U.S. Government.

This study also attempts to objectively measure the effect of European space activities on European identity but finds this difficult due to lack of data. An attempt is also made to assess the effect of the Chinese space programme on Chinese internal cohesiveness. In both cases, the findings are suggestive but by no means conclusive. Finally, it is judged that collective European space efforts are marginally more substantive than Chinese efforts and therefore have the potential to be more effective at shaping group cohesiveness – if they become more visible.

1.1.1 Pride in the group and group identity

An individual's sense of pride in the group to which they belong is a prerequisite for the construction of a strong unified group identity. A synthesis of the rationalist perspective within the social identity field of study finds that an individual's sense of pride in a group is wholly a function of the *amount,* the *perceptibility* and the *duration* of the economic, security, and social benefits, i.e., the material benefits that the group provides to the individual.[7] Group achievements that are perceptible and meaningful in a material sense are more likely to instill a substantive sense of pride in an individual group member. But even though a group achievement is materially worthwhile and perceptible, the pride instilled in group members is not likely to endure if the achievement's material benefits are not provided to individuals on a regular basis, over a long duration. This is because one-time achievements and their material benefits may be more easily overshadowed by other recent group achievements, or failures, or the achievements of other groups.

Even so, there is another crucial ingredient for stiffening an individual's sense of pride in a group – an outside group. The existence of an outside group that individual group members can compare themselves, and their group to, is considered fundamental to the formation of a strong sense of pride in the group. From the social identity school we learn that individuals define themselves and their group in comparison to, and relative to, individuals in other groups.[8] Here it is important to understand the difference between pride and prestige. Pride is the satisfaction that individual group members get from believing that their group has the ability and will to attain worthwhile goals and assure their future. Prestige, on the other hand, consists of a group's *reputation* with other groups for its ability and will to accomplish worthwhile, substantive goals and assure its future.[9] Meaningful symbols of relative differences between groups are key to strengthening pride in the group.

Likewise, meaningful symbols that are perceivable by out-groups affect the in-group's prestige.

Symbols of group achievements are also important because they remind group members of the group's material achievements and they make those achievements perceptible to more group members. However, to be most effective symbols must represent *meaningful* accomplishments, not empty stunts.

1.1.2 Space programmes, pride and prestige, and national identity

Space programmes meet the criteria for providing perceptible, meaningful, material benefits that are enduring and visible to out-groups. However, there are three distinct types of space activities, (1) human spaceflight, (2) robotic space science activities, and (3) space application activities. Each type of space activity generates a different level of material benefit, perceptibility, and durability which are considered below.

1.1.2.1 The material benefits of space programmes

Any type of indigenously developed space programme stimulates the development of a country's industrial, technological, and scientific base, and thereby bolsters development in numerous areas of a national economy. Devices such as launch vehicles (rockets), satellites, sensors, and telecommunications gear are considered the penultimate value-added products. This is often referred to as the "upstream" component of the space sector's economy. Development in these areas requires, in turn, the development of cutting-edge electronics, information technology, telecommunications, materials, project management techniques, and higher education.

Human spaceflight provides these material benefits, and produces important scientific knowledge. However, while human spaceflight does provide significant material benefit to society, it arguably provides less direct benefit to individuals than robotic "space science" activities and so-called "space application" activities. Human spaceflight is also extremely expensive.

In comparison to human space flight, robotic space science activities generate similar "upstream" economic benefits to society, but they deliver more scientific information at much less cost.

Also in comparison to human spaceflight, space application activities such as communication, Earth observation, navigation, and intelligence satellites provide equivalent upstream economic benefits but they also contribute significantly to science, and deliver considerable material and economic benefits directly to society at a reasonable cost.[10] Indeed, without space application systems a modern, "information-based" economy would not be possible as currently construed. All major economic powers recognise this aspect of space and have indigenous space programmes with space application satellites. The U.S. economy and military depend on application satellites so much so that the unhindered use of U.S. space application satellites is considered a vital national interest.

1.1.2.2 The perceptibility of the benefits provided by space programmes

In order for a material benefit to instill pride in an individual for the group, it must be perceivable. Are the material benefits outlined above sufficiently *perceivable* to affect an individual's pride in the group? The material benefits of the three types of space activities are in large part invisible to the general public. Apart from intermittent Space Shuttle launches, occasional images of other planets, and satellite weather loops on the TV, space activities remain outside the public eye. Human spaceflight is the most visible space activity but it provides less meaningful material benefit to the group than the other two types of programmes. The material benefit is largely in the upstream component of the economic realm. In Matrix A, (Fig. 1.1) human spaceflight is subjectively placed toward the "More Material benefit" end of the material benefit continuum but with less material benefit than the other space activities. On the perceptibility continuum, human spaceflight is placed furthest toward the "More Perceptible" end of the scale.

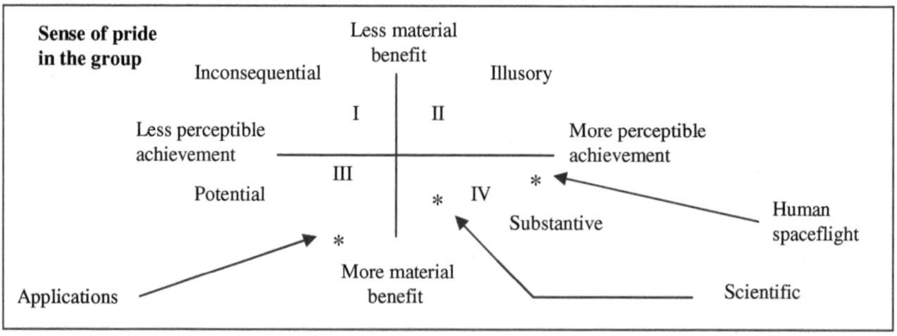

Fig. 1.1. *Matrix A: Sense of pride in the group based upon material benefit and perceptibility.*

With the exception of occasional dramatic images from other worlds, space science activities are much less visible than human spaceflight programmes. In comparison to human spaceflight, robotic space science activities deliver more useful scientific information at less cost while creating equivalent upstream economic benefits. But the direct material benefit to society's members is difficult to discern, especially among the general public. Robotic space science activities are judged to be a bit closer to the "More Material" benefit end of the scale than human spaceflight, but they are placed closer to the "Less Perceptible" end of the perceptibility continuum.

Space application activities provide significant material benefit to society but most of the benefit is absorbed into other products or networks and its presence is transparent to most group members. Hence, space application activities are placed in Quadrant III, closer to the "More Material" benefit end of the scale, but also closer to the "Less Perceptible" end of the perceptibility continuum. Since they provide more material benefit, it is reasonable to argue that space application activities have the potential to foster pride much more effectively than human spaceflight, if they can ever compete with human spaceflight in perceptibility.

That being said, the perceptibility of the material benefits space programmes provide are highly reliant upon symbols. Space programmes provide dramatic symbols that are forward looking, rooted in the future, and based upon solid accomplishments.[11] They are not based upon empty stunts. They generate pride and prestige because they represent meaningful, technological, industrial and scientific accomplishments and because they indicate to the public national purpose, energy and forward momentum.[12] Nevertheless, the symbols mainly derive from the human spaceflight programme, rather than from the more materially beneficial space application satellites. For space programmes to be most effective at strengthening an individual's pride in the group and reinforcing group identity, over the long term, the symbols used to connote space programmes should focus on application satellites and the material benefits they provide rather than on relatively less beneficial human spaceflight programmes.

1.1.2.3 The duration of the benefits provided by space programmes

Assuming the benefits of a space programme are perceivable, its ability to provide enduring benefits is also crucial for its ability to instill a sense of pride for the group in the individual. Figure 1.2 illustrates where each type of space activity falls within Matrix B. Human spaceflight falls within Quadrant III, "Transient," since human

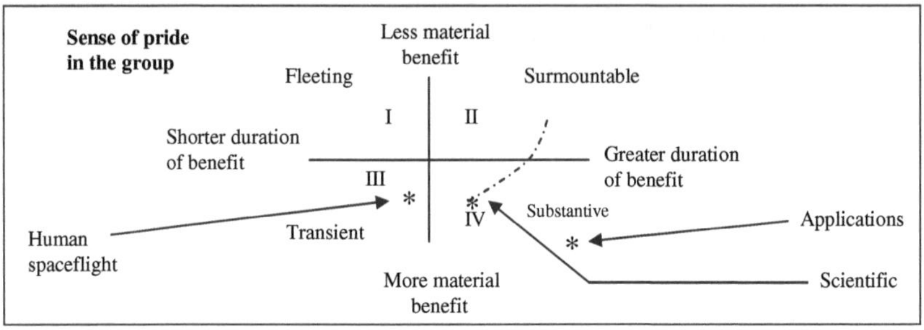

Fig. 1.2. *Matrix B: Sense of pride in the group based upon perceptible material benefit and duration.*

spaceflight produces relatively short-term benefits, other than its effect on the upstream economy.

Scientific space programmes may endure over many years, for example the 33 year old, on-going Voyager I deep space probe and various Mars landers. Nevertheless, the direct meaningful material benefit of such missions goes mainly to scientists and astronomers and eventually tapers off. Thus robotic space science with its longer duration than human spaceflight, but its eroding material benefit in the eyes of the public moves from quadrant IV to quadrant II over time.

Space application satellite programmes, however, continuously deliver a high level of direct material benefit, over long durations. As opposed to their rather weak perceptibility as reflected in Matrix A, in Matrix B they are placed in quadrant IV near the positive ends of both the material benefit and duration continuums.

It is reasonable to argue that space programmes provide identity entrepreneurs an effective means to strengthen group identity. However, each of the three types of space activities has its strengths and weaknesses in this regard. Space application activities deliver the most material benefit to the widest number of people over the longest time. However, to be effective at bolstering group identity they need to be more perceptible. Human spaceflight, on the other hand, fosters pride mainly because it is more visible, even though it delivers less material benefit, is not cost effective, and is relatively not very enduring.

1.1.2.4 Collective European space programmes, pride, and European identity

European identity entrepreneurs argue that the European identity must be forward looking, rooted in the future, and use concrete symbols.[13] EU space activities support these criteria very well.

Today, EU space activities are focused on using space application satellites to deliver a high level of direct, meaningful material benefits to the European population. The main effort of EU space activities is to place in orbit navigation and Earth observation satellite systems. While these systems have not been launched yet, they are on the threshold. United Press International (UPI) describes such programmes as embodying the "stunning potential of a unified Europe".[14] Their weakness, at present, is their lack of perceptibility. The *International Herald Tribune* reports that Europe's "startlingly ambitious" achievements in robotic space science missions to the planets are "relatively unsung" and are almost entirely unknown to the public.[15] The same may be said about European space application achievements. Thus assessing European public opinion for the sense of pride engendered by space activities, and their effect on EU identity, should reveal that when Europeans are cognizant of their space achievements, they develop a substantive sense of pride and hence a greater sense of EU identity.

Public opinion survey information on European attitudes toward European space activities is difficult to find. However, some insights may be developed by using "science and technology" as a proxy for space. Useful information may be derived by using Eurobarometer (EB) survey data on European attitudes toward science and technology and synthesising it with other EB statistics that show that the sense of European identity is growing, especially among younger people.[16] The June 2005 EB survey, *Europeans, Science and Technology* indicates that there is a positive perception of what science and technology in Europe can do for the quality of life. Furthermore, in 2005, 23% of respondents who said they were "very" or "moderately" interested in science and technology identified "astronomy and space" as the category of science and technology in which they were most interested. This is up from 17% in 2001, the biggest rise in interest among the science and technology categories. The EB survey says this is likely explained by the media attention European space achievements have received.[17] Interestingly, those age 24 and younger show more interest in science and technology, 33% and 38% respectively, than the EU average of 30%.[18] This is not enough data to say confidently that interest in space among European youth correlates to a strengthening of a European sense of identity among youth, but it is suggestive. Also, the data supports the idea that the perceptibility of space achievements appears to be a key factor in their usefulness for strengthening group identity. It is reasonable to argue that for the EU space programme to be most effective at promoting a stronger European identity, EU space activities should be more visible. As well, powerful symbols of EU space activities must be developed that stimulate a sense of pride in EU citizens by reminding them of the material benefits the EU space programme provides to them individually and as compared to individuals in other groups.

1.1.2.5 The Chinese space programme, pride and Chinese identity

China has a variety of ethnic groups and the cohesiveness of the Chinese State cannot be taken for granted. Accordingly, *The Economist* reports that instilling national pride to help the country remain united is an important motive behind China's human spaceflight programme.[19] James Oberg argues that the Chinese space programme's contribution to national pride cannot be "sneezed away" as a genuine benefit.[20] He points out that national unity has been a theme throughout Chinese history and notes that without "the mandate from heaven" the central government can fall. Oberg notes how evocative a symbol the Chinese space programme presents in this regard.[21] In addition, Dean Cheng, a U.S. expert on the Chinese space programme, notes that there is a lot of unrest in the Chinese underclass, and a successful human spaceflight programme "could paper over the cracks for a while".[22] Thus it is not surprising that the 2006 Chinese White Paper "China's Space Activities in 2006," states that one of the objectives of China's space programme is to be "a cohesive force for the unity of the Chinese people".[23]

The weight of the Chinese effort is behind its human spaceflight programme, with ten times more spending on it than on space science and space application satellite programmes.[24] According to the framework offered above, perceptibility is its main strength. National Public Radio, reporting on a Chinese human spaceflight stated, "Perhaps no payoff is as immediate as the public relations one. Not many images can compare with the blast-off of the manned spacecraft for summing up a nation's economic and scientific capabilities".[25] *Time International* reported that China's first human spaceflight in October 2003 triggered a "historic outpouring of national pride," with the State controlled media extolling the accomplishment's material benefits to science, technology, and collective will. However, in the same sentence, *Time* noted that the flight was only 21 hours long.[26]

According to the reasoning outlined above, the first potential weakness of the Chinese space programme is that its material benefits belong primarily to the upstream economic arena. Second, the comparatively short duration of human space flight activities means the pride fostered is probably transient in nature. On balance, it appears as if the Chinese space programme, with its focus on human spaceflight, will generate significant levels of pride, but not be as effective as possible in promoting a stronger Chinese identity, over the long term.

Based on the analysis presented above, there should be indicators that show the sense of pride engendered by the Chinese space programme is significant, but not highly substantive or enduring. However, with the exception of Hong Kong, information is tightly controlled in China and there is no freedom of the press. Meanwhile, the State controlled media plays up the material benefits of human spaceflight while ignoring the cost.[27] Surveys of how the Chinese feel about their space programme are not available. But opinions about the Chinese space programme expressed in the Hong Kong-based Chinese media are suggestive proxy indicators. In addition, online blogs and chat-rooms have become a means for the Chinese to express their opinions, although the data they provide is suspect and non-scientific. But since blogs are monitored by authorities, expressing an opinion contrary to the government line carries some risk.[28] The existence of any contrary opinions in blogs should be given some extra consideration since they imply that the opinion is strongly felt and State provided, force-fed, messages are not completely convincing to the population.

James Hansen, a Pulitzer Prize nominated historian, pointed out that when the first Chinese astronaut visited Hong Kong on a "human spaceflight exhibition," the local media was cynical, and local newspapers and business leaders criticised the visit as propaganda designed to prop up the Communist government. But there was substantial evidence that the locals separated the individual achievement of the astronaut from the propaganda and were excited about the visit.[29] The Hong Kong based *South China Morning Post* reported that most mainland China online chat room contributors supported the space programme as a source of national pride, but a variety of negative opinions also appeared. The negative comments were characterised mostly as complaints that the money invested would be better spent on more directly beneficial material goods.[30] A week later the *South China Morning Post* published a letter that expressed the same ideas stating, "While millions of our compatriots suffer from poverty, it is not national glory to burn money on a rather meaningless space trip".[31]

The above analysis provides some support to the argument that the level of material benefit that space achievements deliver to group members, and the duration of the benefit, affects the usefulness of the programme for strengthening group identity. It is reasonable to argue that in order for the Chinese space programme to be most effective at promoting a stronger Chinese identity over the long term, the Chinese should ensure their human spaceflight programme is balanced with significant, well publicised space science and space application endeavors. Otherwise, too much focus on one-time or short term space "events" may eventually be perceived as empty stunts by the Chinese population, and become a symbol of wastefulness or misplaced priorities.

1.1.3 Conclusion

This study places the claim that space programmes strengthen group identity on firmer ground by shedding light on the relationship between space programmes, pride, and identity. It argues that the ability of space programmes to strengthen national identity is a function of the perceptibility, duration, and meaningfulness of the material benefits the programme provides. An examination of Chinese space activities and collective European space activities suggests that European space efforts are marginally more meaningful than Chinese efforts and therefore have the potential to be more effective at shaping group cohesiveness – if they become more visible. This study also points out the direction for further research into the relationship between pride and prestige, and group identity. An effort to develop empirical evidence through scientific surveys that specifically target opinion about space programmes and their effect on pride and identity would be a useful next step.

[1] Oberg, James. Space Power Theory. From a draft by Dr. Brian R. Sullivan, p. 136.

[2] McDougall, Walter. The Heavens and the Earth: A Political History of the Space Age. Baltimore and London: John Hopkins University Press 1985, p. 390.

[3] Oberg, James. Space Power Theory. From a draft by Dr. Brian R. Sullivan: 169.

[4] Commission of the European Communities. Communication from the Commission to the Council and the European Parliament: European Space Policy. COM (2007) 212 of 26 April 2007. Brussels: European Union.

[5] Information Office of the State Council, China's Space Activities in 2006, 12 October 2006.

[6] Rusciano, Frank. The Construction of National Identity: A 23-Nation Study. Political Science Quarterly 56 (2003): 361.

[7] Posner, Daniel. Institutions and Ethnic Politics in Africa. New York: Cambridge University Press 2005, p. 11.; Katab, George. The Inner Ocean, referenced in: Mayerfeld, Jamie. The Myth of Benign Group Identity: A Critique of Liberal Nationalism. Polity 30 (1998): 555.; Rusciano, Frank. The Construction of National Identity: A 23-Nation Study. Political Science Quarterly 56 (2003): 361.; and Jamie, The Myth of Benign Group Identity: A Critique of Liberal Nationalism. Polity 30 (1998).; Horowitz, Donald. Ethnic Boundaries, Riot Boundaries, The Deadly Ethnic Riot. Berkley: University of California Press 2001 p. 47.; Isenberg, Arnold. "Natural Pride and Natural Shame." Philosophy and Phenomenological Research, 10.1 (1949): 1.; Anderson, Benedict. Imagined Communities. London: Verso 1983.

[8] Ashforth, Blake E. and Fred Mael. "Social Identity and the Organization." The Academy of Management Review 14.1. (1989): 21, citing Tajfel and Turner, 1985, p. 16.

[9] Frye, Alton. "Politics–The First Dimension of Space." The Journal for Conflict Resolution 10.1 (1966): 105.; review article of Vernon Van Dyke, Pride and Power: The Rationale of the Space Program. Urbana: University of Illinois Press 1964.

[10] The material and economic benefits that flow from such space application activities are often referred to as the "downstream" space sector and produce more revenue than the aforementioned upstream space sector.

[11] I owe these ideas to Peter Van Ham in International Politics cited in: Stutzmann, Alexandre. "Europe's Fake ID." Foreign Policy 126 (2001): 94.

[12] Frutkin, Arnold. "The United States Space Program and Its International Significance." Annals of American Academy of Political and Social Science 366 (1966): 89.

[13] Stutzmann, Alexandre. "Europe's Fake ID." Foreign Policy 126 (2001): 94.; Kaelberer, Matthias. "The Euro and European Identity: Symbols, Power, and the Politics of European Monetary Union." Review of International Studies 30 (2004): 161–178.
[14] "Europe Boosted by Reactor, Galileo." 28 June 2005. UPI, 15 December 2006. [database on-line], Lexis-Nexis.
[15] "Don't Look Now, but Europe is in Outer Space." International Herald Tribune, 21 July 2005: 6.
[16] "European Identity on the Rise, Says Austrian Study." 17 November 2006. Austria Today 15 December 2006. [database on-line], Lexis-Nexis.
[17] Commission of the European Communities/Eurobarometer., "Europeans, Science and Technology." Special Eurobarometer 224. June 2005: 15.
[18] *Ibid.*: 8.
[19] "Pride and Prejudice," 25 March 2006. The Economist, 15 December 2006. [database on-line], Lexis-Nexis.
[20] Oberg, James. "The 'Why' Behind China's Own Private Space Race." USA Today, 17 October 2005.
[21] *Ibid.*
[22] Aldhous, Peter and Anil Ananthaswamy. "Asia Blazes a Trail to the Final Frontier." New Scientist 88.2522 (2005): 8.
[23] Information Office of the State Council, China's Space Activities in 2006, 12 October 2006.
[24] "Is Space Travel Worth the Money?" 18 October 2005. South China Morning Post, 15 December 2006. [database on-line], Lexis-Nexis.
[25] Kuhn, Anthony. "China Launches Manned spacecraft." National Public Radio Morning Edition, 12 October 2005.
[26] Frederick, Jim "Asia's Space Race," 10 October 2005. Time.com, 21 November 2010. *http://www.time.com/time/magazine/article/0,9171,1115727,00.html.*
[27] *Ibid* (Time International).
[28] Reuters. "Battle of the Blogs in China." 10 August 2005. International Herald Tribune.
[29] Day, Dwayne A. "Exploring the Social Frontiers of Spaceflight." Space Review, 25 September 2006.; Summary of presentation given by James Hanson at Smithsonian National Air and Space Museum at "Societal Impact of Spaceflight." conference, 19–21 September 2005.
[30] "Is Space Travel Worth the Money?" 18 October 2005. South China Morning Post, 15 December 2006. [database on-line], Lexis-Nexis.
[31] "Instead of Space Mission, Help Mainland's Poor." South China Morning Post, 25 October 2005: 16.

International relations

1.2 The European identity in space: an international relations perspective[32]

Christophe Venet

1.2.1 Introduction

Most contributions in the present book analyse the link between a European identity and the European Space Policy (ESP) from a European domestic point of view. The underlying rationales of these reflections are mostly rooted in internal factors: how does a *specifically European* identity influence a *specifically European* space policy and vice versa? This contribution intends to add an international perspective to these thoughts. As such, it will tackle the external dimension of the European identity in space, focusing on the behaviour of Europe on the international scene and on the perception of the ESP by other spacefaring nations.

The central goal of this article is to find out if and how the European foreign policy identity debate can be applied to the ESP. A first important point to make in this respect is that the ESP is indeed part of the broader European foreign policy. Foreign policy can be defined as the "endeavour of a particular actor, usually a state or, more rarely, a union of states, to protect its interests and to promote its values vis-à-vis other actors beyond its borders."[33] A first assumption that can be made for the purpose of this paper is that the EU can be considered as the main political space actor in Europe (a "union of states" according to the above-mentioned definition).[34] From a concrete perspective, the EU is already integrating science and technology (S&T) cooperation endeavours, including space-related ones, into its foreign policies. As such, it makes sense to talk about the "EU's emergent space diplomacy".[35] Furthermore, the Resolution on the ESP underlined the role international cooperation can have in "strengthening Europe's role in the global space field".[36] As a consequence, a document called "Elements for a European Strategy for International Relations in Space" was adopted in 2008, highlighting that such a strategy should be in line with the EU's external policy priorities.[37] As a whole, the ESP has a clear external dimension, and can thus be considered as part of the EU's foreign policy. However, an important question in this respect is to know whether this is a conscious and politically motivated strategic decision, or if the ESP has been developed independently without a clear link to the broader EU foreign policy objectives.

Since the foundation of the European Communities, one of the most debated discussions among scholars of international relations focuses on Europe's foreign policy. Two central questions are at the core of these research endeavours:

1. Is the EU an international actor?
2. Is there a specific European international identity (i.e., what sort of international actor is the EU)?

These two questions are intrinsically linked to each other. Indeed, the existence of a specifically European identity on the international scene presupposes the existence of an identified actor to which such an identity can be attributed (i.e., the EU). The aim of this paper is to transpose this debate to the specific issue-area of space.

The question of Europe's identity in space also touches upon broader questions discussed in international relations. In particular, a reflection on Europe's identity at the international level implicitly contains a reflection on the relative weight of structures and agents, and on their interaction. Those are the terms of the agency-structure debate, popularised by Alexander Wendt,[38] asking whether actors (states) can shape their environment (the international structures) or whether their scope of action is limited by structural constraints. This is another way to look at the question of international identities: are domestic features and specificities (i.e., an actor's identity) relevant on the international scene, or is any functional difference among actors impossible?

It has to be noted that this contribution only intends to take a first look at these issues. It will therefore lay down assumptions, but will in no way lead to conclusive evidence, thus leaving room for further investigation.

To answer the above-mentioned questions, it is first necessary to briefly look at the scholarly literature on EU's international actorness and on the international identity of the EU. In a second step, the specific methodological difficulties of applying the debate to the space sector will be highlighted. Finally, the actorness and international identity of the EU in space will be analysed, by showing, respectively, that the EU is a growing actor in space, and that its identity can be assessed as a mix between norms and interests.

1.2.2 The scholarly debates

Chronologically, the literature on EU's international actorness has focused on two issues: first, the fundamental question of whether the EU is an international actor at all, second the question of the characterisation of the EU's foreign policy.[39]

1.2.2.1 The EU's international actorness

Initially, realist scholars came to the abrupt conclusion that the European Community/the European Union is not an international actor. These analyses mostly focused on the lack of military power characterising Europe's foreign policy, while military capabilities were understood as being a central element of international actorness.[40] However, some actor-like features were progressively attributed to the EC as a non-state actor in the 1970s. This mainly resulted from the growing popularity of the concept of "complex interdependence".[41] A series of empirical studies were conducted, showing that Europe could indeed be considered an international actor in certain policy areas.[42] However, a common feature of these analyses was a shift of focus from "high politics", meaning security and defence policies, towards "softer" policy areas, such as trade relations, humanitarian relief or neighbourhood policies.[43] More recently, the Lisbon Treaty marked an important step forward by creating a single legal personality for the EU. While it may take time for this to be followed by concrete foreign policy realisations, as testified by the difficulties encountered in setting up the new European External Action Service (EEAS), it nevertheless paves the way for the EU to play an enhanced role on the international scene.

1.2.2.2 The EU's international identity

The question of the EU's international identity is discussed along a line between two opposite poles. On the one end are liberal-idealistic arguments, presenting the EU as a civilian and normative power which behaves differently than other international actors, on the other end of the spectrum are realist authors, highlighting the structural constraints of the international system excluding any normative specificity of the EU.[44]

François Duchêne was the first to conceptualise Europe as a civilian power. According to his analysis, Europe would use civilian means to reach civilian goals, resting on social values such as tolerance, justice and fairness.[45] The end of the Cold War led to a renewed interest in the normative aspects of European foreign policy, against the background of the so-called "reflexive turn" in international relations.[46] In the early 1990s Maull further developed the concept of civilian power using the examples of Germany and Japan.[47] Manners later introduced the concept of the EU as a normative power, trying to shift away from the military/civilian power debate to focus instead on the "ideological power" of the EU. Accordingly, the EU put norms at the core of its relations both with its Member States and with third States, through the combination of a particular historical context, a hybrid polity and its legal constitution.[48]

The neo-realist critiques of these approaches recalled the structural factors of the international system, namely its inherent anarchy (i.e. the absence of a regulating body with a monopoly on legitimate violence at the global level) and the unequal power distribution among states. According to neo-realists, liberal-idealistic perspectives both ignore these "hard" facts and base their arguments on normative preconceptions.[49] To the contrary, the foreign policy behaviour of the EU can be best explained by systemic pressure. The development of the European Security and Defence Policy (ESDP) for example, can be understood as a classical balancing policy against the U.S.[50] Liberal-idealistic literature in turn, countered these arguments with the concept of "civilising power". The EU foreign policy would not only be based on norms, but a major foreign policy objective of the EU would be to transform power structures by disseminating its norms to the outside.[51] However, a danger that was also recognised in the liberal-idealistic literature is the potential discrepancy between the self-assertion of the EU as a normative actor and the actual realities of international politics.[52]

After this quick overview of the scholarly debates, it becomes clear that Europe's international actorness and identity varies strongly depending on the policy-area considered. While it seems fair to admit that the EU is not only an important but also a distinctive actor in several "soft" domains, it seems more difficult to label it an international actor in defence and security policy.[53]

1.2.3 Methodological difficulties in the case of space

There are two major methodological difficulties in applying the debate on Europe's international identity to the issue area of space. The first concerns the multiplicity of stakeholders in the ESP, making it difficult to identify a single relevant actor. The second is that space is a complex issue-area at the crossroads of several policy areas.

The multiplicity of actors is a classical issue in European foreign policy analysis. Indeed, it is necessary to consider three different levels of analysis:

the Community, the Union, and the Member States.[54] This applies to the governance architecture of the space sector in Europe as well. Many Member States, especially the most active in space, have national space programmes, often coordinated and managed by national space agencies. In addition, two intergovernmental institutions are active in space matters. EUMETSAT is the European operational satellite agency for monitoring weather, climate and the environment. ESA is the oldest and most active European actor in space. Finally, the EU has become increasingly involved in space in the last decade, some responsibilities being held by the European Commission (such as the Galileo and GMES programmes), some other issues being discussed by the Council of the EU (mainly CSDP-relevant space activities). The situation is even further complicated by the fact that the membership of these three institutions (EUMETSAT, ESA and the EU) does not necessarily overlap.

However, there seem to be several good reasons to consider the EU as the main European foreign policy actor in space. Space by itself gained increased political relevance after the Cold War, moving away from a strictly scientific and technological perspective. It seems thus natural that the main political actor in Europe sought to integrate a space dimension into its broader socio-economic policies. Concrete manifestations of this trend included the launch and management by the European Commission of two ambitious programmes – Galileo and GMES – and the recent EU diplomatic initiative of the Code of Conduct (CoC) for Outer Space Activities. ESA maintains a broad array of cooperation ties with foreign space agencies, but most of these are motivated by programmatic priorities and cost-effectiveness concerns rather than by strictly political considerations. The governance structures of both Galileo and GMES reflect the EU's growing role in space, attributing the overall political guidance to the EU, and the technology development and programme implementation aspects to ESA. This division of labour also translates into the international dimension of European space activities. The EC/ESA Framework Agreement of 2003 already contained an article on external relations, calling on the two parties to inform and consult each other on their respective international activities.[55] The main rationale here was the necessity to speak with one voice on the international scene, and the EU seems to be the adequate European institution to do so. Finally, this growing Europeanisation of space policy should be understood as a dual process: not only does European integration lead to domestic changes within Member States, but these domestic changes in turn can also reinforce integration trends.[56] This means concretely that while the EU showed increased interest for space (integration dynamic), Member States also progressively realised that Europe was the only possible level of action to fulfil their space ambitions (domestic change). As a whole, even if the above-mentioned assumptions need to be further investigated to be confirmed, it seems reasonable to consider the EU as the main European international actor in the issue area of space.

Second, space is a complex issue area at the intersection of the three international policy fields identified by the political scientist Czempiel: security, welfare and system of rule.[57] Indeed, space assets can be used for military purposes (security), they can contribute to the stimulation of economic growth, the equal distribution of wealth and benefits produced by the economic growth, and the sustainability of these processes (welfare) and space can finally be a way to affirm a state's sovereignty on the international scene (system of rule). As such, space policy touches upon both "high politics" and "soft" areas. As described above, the actorness and international identity of the EU varies greatly from one policy area to the other. As a consequence, it seems necessary to take this into account in the assessment of the European identity in space. More specifically, a distinction has to be made between Europe's space security policies and the EU's civilian international policies in space.

1.2.4 Europe is a rising international actor in space

As described above, a first step towards the characterisation of Europe's identity in space is to find out whether the EU is an international actor in space. To do so, it seems necessary to define a framework for analysis containing a set of observable criteria to assess the EU's international actorness. Sjøstedt was the first to conceptualise actorness by identifying two structural conditions: commonly accepted goals and the existence of a system for mobilising resources necessary to meet the goal.[58] The concept was further refined by the introduction of new constitutive elements: "actor capacities"[59] and "presence".[60] The model developed by Bretherton and Vogler finally, integrates these developments to offer a dynamic picture of the EU's actorness taking into account the uniqueness of the EU as an international actor and the complex interaction between structures and agencies. The two scholars identified three central features to focus on: "opportunity", "presence" and "capabilities".[61] In the following, these three concepts will be briefly sketched and applied to the issue area of space.

"Opportunity" first, "denotes factors in the external environment of ideas and events which constrain or enable actorness. Opportunity signifies the structural context of action."[62] In this regard, both material conditions and inter-subjective structures are taken into account. As a general trend, space gained an increasing political nature after the end of the Cold War, moving away from a narrow S&T driven field. This led to a growing number of countries engaging in space activities (in particular developing countries) and in the multiplication of regional and global space-related cooperation endeavours. As a result, opportunities for EU foreign policy activities in space multiplied. Another important factor stimulating international space cooperation is the growing consciousness that space assets can be used to solve global transnational problems. It is thus necessary to tackle these challenges with interdependent and globalised solutions. A good example in this case would be the contribution of GMES to the fight against climate change. Finally, a strong material driver behind most international interactions in space is the need to pool limited national financial and technological resources to engage in more ambitious programmes, the ISS epitomising this phenomena. These combined structural factors create a favourable environment for the EU's international engagement in space, based on a demand for political action mostly in the form of international cooperation. In the field of space security, on the contrary, cooperation seems to represent a more distant prospect. However, there are equally strong structural factors that would call for EU action at the international level. The material structures of the international system, based on anarchy and power competition, lead to a security dilemma.[63] The 2006 U.S. National Space Policy, clearly stating the possibility to "deny, if necessary, adversaries the use of space capabilities hostile to U.S. national interest",[64] the development of anti-satellite (ASAT) capabilities by China or India, and the long-lasting stalemate at the Conference on Disarmament (CD) are some of the most striking manifestations of this security dilemma. On the other hand, the discrepancy between European aspirations in the field of defence and security and its actual capabilities – the famous capability-expectation gap – is a factor that could constrain European action in space security. As a whole, it seems that the structural environment is rather an enabler than an obstacle to the EU's international actorness in space, both in civilian and military space.

The second criterion, "presence", refers to "the ability to exert influence externally; to shape the perceptions, expectations and behaviours of others."[65] The concrete question here is to determine the reputation and status of the EU as perceived by external actors, which can be measured by three indicators. The first one is the character of the EU, being understood as the EU's material existence (i.e. its political system). In the case of space, external actors can identify their European interlocutors easily. ESA is the sole pan-European partner for other space agencies around the world wishing to cooperate on a programmatic basis (for example ESA/NASA cooperation on Mars exploration missions). The European Commission (EC) then, acting on behalf of the EU, deals with the political aspects of space foreign policy. It is the EC, for example, which negotiates and signs the Galileo compatibility and interoperability agreements. In the area of space security however, the situation is not that clear. Although the EU starts to emerge as a diplomatic actor in this field (presenting and defending the draft CoC on the international scene), there is no integrated European space security architecture yet. As a consequence, in this field the institutional picture given by the EU to the outside is still blurred. The second indicator, identity, refers to "shared understandings that give meanings to what the EU is and what it does."[66] This aspect is very complex, as identity is both a result and a component of the EU's actorness. The EU's identity can be conceived of as a mixture of both inclusive (the EU as a value-based community) and exclusive (excluding participation of the "other") factors.[67] Europe's international activities in space seem to reflect this duality: while the EU is highlighting cooperation and multilateralism as central norms in all its space-related official documents, it pursues a more exclusive policy in other aspects, like industrial policy (non-European firms were, for example, excluded from the Galileo industrial process). Finally, the last indicator of "presence" describes the external consequences of EU actions, which are often unintended or unanticipated. Examples in space could include the reactions of other states to the two European flagship programmes Galileo and GMES. While certain states perceived these systems as potential competitors (the other GNSS providers in particular), their positive potential was also acknowledged (for example the role GMES could play in the Russian International Global Aerospace Monitoring System (IGMASS) proposal[68]). The perception of the EU as a space security actor by other states seems more difficult to establish. The main reason for this is the lack of European initiatives in this field so far. However, the introduction of the draft CoC in international fora was accompanied by intensive consultations with other spacefaring nations, resulting in de facto interaction between the EU and these other actors. Similarly, certain voices have started to call for a consideration of the EU as a space security actor, in particular in the U.S.[69]

The third and last criterion, "capability" refers to the "internal context of EU action or inaction – those aspects of the EU policy process which constrain or enable external action and hence govern the Union's ability to capitalize on presence or respond to opportunity."[70] Four indicators can help to operationalise the "capability" criterion.[71] First, shared commitment to a set of overarching values is needed. This seems to be given in the case of space: while the broader framework is given by the values entrenched in the Treaty on European Union (TUE),[72] the Council resolution on the ESP recalled that Europe's space activities should be based on the principles set out in the Outer Space Treaty (exploration and use of outer space for the benefit of all mankind, peaceful purposes, promotion of international cooperation).[73] In the field of space security, these principles could be supplemented by the elements laid down in the European Security Strategy (ESS), such as a commitment to multilateralism, the European readiness to take more international responsibilities or a broad understanding of security.[74] The second requirement, domestic legitimation, is more problematic. While being a structural weakness of the EU as a whole, the democratic deficit is particularly striking in the ESP.[75] The third indicator is the ability to identify priorities and to formulate policies. In this regard, a central condition is the consistency between national Member States and EU policies. The ESP is implemented by an institutional framework regulating the contribution of all the stakeholders to the ESP (including the EU and the Member States), and aiming precisely at avoiding unnecessary duplications. The main decision-making body in this respect is the Space Council, set up through the EC/ESA Framework Agreement of 2003. On the other hand, the shared competence on space introduced by the Lisbon Treaty might take time to materialise, given a latent reluctance among Member States to go too far in the Europeanisation of the ESP. In space security, consistency is virtually non-existent, as the vast majority of initiatives and military operational capabilities are in the hands of the Member States. The coherence of the ESP with other EU policy fields seems also to be progressing. For example, space has been recognised as an important contribution to the Europe 2020 strategy,[76] and more generally, a central policy objective of the ESP is to integrate space in the socio-economic policies of the EU. However, the lack of a systematised and structural link between the ESP and the general foreign policy objectives of the EU is still problematic. It seems indeed paradoxical to highlight the contribution of space to EU policies without articulating a clear space foreign policy. The last indicator is the availability of policy instruments in the form of political (diplomacy/negotiations), economic and military means. While these conditions seem to be given in the case of the civilian aspects of the ESP, the lack of political, economic and military instruments at the EU level certainly has an impact on the underdeveloped use of space assets for security at a European level.

To summarise, it seems fair to say that the EU is indeed an international actor in space policy, although an incomplete one. A clear differentiation has to be made between civilian space and military aspects. Indeed, given the sensitive nature of military space activities and the peculiar nature of the European polity, the EU is only in its infancy in this field, and although it is striving for international actorness, there seem to be a long way ahead.

1.2.5 Europe's international identity in space

As the EU clearly has attributes of an international actor in space, a second question is to find out whether it is also a *distinctive* actor, meaning if its specific identity has an influence on the way she behaves on the international scene. To find out, one can apply two international relations theories, neo-realism and social constructivism. Neo-realist explanations focus on the European striving for strategic autonomy, which is a logical consequence of structural constraints and make the EU an international actor no different from other spacefaring nations. On the other hand, constructivist analysis puts emphasis on the role of norms in shaping the EU foreign policy, thus enabling a differentiation from other actors.

In central space policy documents, the EU proclaims its ambitions to become a leading space power at the global level.[77] As stated in the ESP, the EU should be ready to "assume global responsibilities" and an "effective space policy would enable Europe to exert global leadership in selected policy areas". Similarly, Europe should "secure unrestricted access to new and critical technologies, systems and capabilities in order to ensure independent European space applications."[78] These strategic orientations can be analysed as a reaction to changing structural features. These include not only the rise of ambitious new space actors,[79] but also the development of a U.S. space strategy aiming at "full spectrum dominance" under the Bush administration,[80] although these evolutions were tamed by the new orientation given to U.S. space policy by the Obama administration. In this overall context, some authors call it rational for Europe to develop asymmetric means to increase its space capabilities.[81]

The neo-realist approach in international relations is fitted to theoretically capture and explain these features of the ESP. A basic neo-realist assumption is that states follow their preferences rationally while their ultimate objective is to survive, but each state in the international system has different power capabilities. The structure of the international system in turn, has a direct causal influence on its units, i.e. states. It is composed of three elements: the ordering principle (anarchy or hierarchy), the character of the units (existence or absence of functional differences between states) and the distribution of capabilities (unipolarity, bipolarity or multipolarity). Neo-realists consider the international system to be anarchical, which in turn forces every single state to pursue self-help strategies to ensure its survival: functional differences between actors of the international system are impossible.[82] While the distribution of capabilities in the current international system is still a debated issue, states have to make sure in any case to keep a power balance in the system.

Translated to the European foreign policy in the issue area of space, this would mean that the EU is pursuing an autonomy maximising policy following the worsening of its relative power position in the system.[83] A certain number of elements seem to validate this approach. Among the structural factors endangering the EU's power position are the above-mentioned rise of new spacefaring actors, the prospects of an arms race in outer space, fuelled by an enduring stalemate at the

Conference on Disarmament, and the decline of Europe's capabilities in space (space budgets for example), relative to other states. As a consequence, the rationale of strategic independence has always been highlighted by Europe in its space endeavours. It started with the launch of the Ariane launcher programme to gain independent access to space, and continued with the Galileo and GMES flagship, both destined to ensure Europe's strategic autonomy in the key applications of Earth Observation (EO) and Global Navigation Satellite System (GNSS). Recent EU initiatives to participate in the Space Situational Awareness (SSA) programmes and to strive for European non-dependence in certain critical technologies are also going in the same direction. As a whole, this seems to show that the EU is behaving not differently than any other spacefaring nation, trying to protect its interests and to keep its power position in the international system. However, certain steps made by the EU in space recently seem to indicate that there are also aspects other than structural constraints to be taken into account when analysing the EU's international identity in space.

A striking feature in official EU documents related to space is the constant reference to the international norms enshrined in the space treaties (in particular the peaceful uses of outer space) as well as to specific European values.[84] Social constructivist approaches precisely focus on the independent influence of norms in international relations. According to the neo-realist paradigm, actors maximise their utility by following a logic of consequentiality. To the contrary, social constructivists argue that actors follow a logic of appropriateness: they take norm-oriented decisions on the background of their historic and cultural experiences, and their institutional ties.[85] A central idea of the constructivist research programme is the reciprocal influence of actors and structures. In this perspective, it is necessary to look at the influence on foreign policy of both international and domestic norms. To measure the impact of a norm, both its commonality (i.e. the number of actors in a given social system sharing the same behavioural expectations induced by the norm) and its specificity (i.e., the precise distinction between appropriate and inappropriate behaviour) have to be assessed.[86] Concretely, to assess whether the EU behaves differently than other actors in space, one has to first identify the specific norms associated with space policy, both at the international level and domestic EU level, and then to find out if the EU's behaviour on the international scene is consistent with these norms.[87]

A series of specific norms is entrenched in the space-related international treaties.[88] Among the most relevant ones are the exploration and use of outer space for peaceful purposes and for the benefit and the interest of all countries, the prohibition of any kind of national appropriation in outer space and the right of each single state to explore outer space. These basic principles are systematically recalled in any official space policy document, be it at national or international level. The EU is no exception in this respect and both the Council and the European Parliament keep highlighting the importance of these principles in their documents.[89] At the EU domestic level then, Manners identified five specific norms constituting the normative basis of the EU: peace, liberty, democracy, rule of law and human rights.[90] In international space policy, only peace seems to constitute an applicable principle. A concrete translation of this peaceful approach could be found both in the emphasis given by the EU on multilateralism and international cooperation in space, as well as in the civilian primacy in dual-use programmes. A whole chapter was, for example, dedicated in the latest Space Council resolution to the partnership on space with Africa.[91] Similarly, GMES and Galileo are officially dubbed civilian programmes under civilian control, despite their obvious dual-use potential. The last noteworthy EU step in this framework is the draft Code of Conduct for outer space activities proposed to the other spacefaring nations and aimed at overcoming the deadlock at the CD. These initiatives seem to give some credit to the thesis of the EU as a "civilising power", as the EU aims at transforming the parameters of power politics both through the strengthening of the international legal system, and through the externalisation of its domestic norms.

1.2.6 Conclusion

Although this contribution provided only a first look at the question of the EU's international identity in space from an international relations perspective, a few elements may be highlighted. As a starting point, it seems necessary to make a distinction between civilian and military space activities when assessing Europe's identity in space. Europe clearly established itself as an international actor in civilian space in the course of the last decades, through ESA first and now with the EU. In military space, this trend is less obvious, as Europe is a very young actor in this field, and is still struggling to set up a coherent space security strategy. Similarly, it is both more coherent and more difficult to assess Europe's specific identity in military space than in civilian space activities, as military space is more value-loaded than civilian space activities. Here again, it might be too early to make a definitive and clear assessment about Europe's normative distinctiveness.

Summarising, it seems that Europe's space policy is shaped both by structural factors (the constraints of the anarchical international system but also the transnational norms laid down in the space treaties), and by domestic factors (the specific features making up a distinctive European identity). In this sense, it might be too far-fetched to consider Europe as a radically different actor: like the other spacefaring nations, Europe is following «realist» goals, such as the maximising of its autonomy. On the other hand, empirical evidence also suggests that Europe is more prone than other actors to putting emphasis on norms in its space policy. Identity in this respect does not directly determine interest, but it still has a "mediating function".[92] Thus, it seems possible to apply Smith's argument of the EU as a hybrid of civilian power and power bloc[93] to the issue area of space. This is perfectly laid down in the ESP: "Europe needs an effective space policy to enable it to exert global leadership in selected policy areas *in accordance with European interests and values.*"[94] This European specificity in space-related foreign policy is starting to be recognised by other spacefaring actors, as witnessed by the serious interest shown by the international community – including the U.S. – in the EU proposal of a Code of Conduct. Finally, the rising international profile of the EU in space matters could also have a stimulating effect on the internal construction of a European identity. As such, Europe's space foreign policy would both rest on elements of a European identity and strengthen in return the European identity as a whole.

---

[32] This article is partly based on the research design of the author's PhD project.

[33] Rittberger, Volker. Approaches to the Study of Foreign Policy Derived from International Relations Theories. Tübinger Arbeitspapiere zur Internationalen Politik und Friedensforschung Nr. 46. Tübingen: University of Tübingen 2004: 1.

[34] This argument will be further developed in Sect. 2.

[35] Peter, Nicolas. "The EU's emergent space diplomacy." Space Policy 23:2 (2007): 97–107.

[36] Council of the European Union. Resolution on the European Space Policy. Doc. 10037/07 of 25 May 2007. Brussels: European Union.

[37] Commission of the European Communities. Commission Working Paper. European Space Policy Progress Report. COM (2008) 561 final of 11 September 2008. Brussels: European Union.

[38] Wendt, Alexander E. "The agent-structure problem in international relations theory." International Organization 41:3 (1987): 335–370.

[39] Ginsberg, Roy H. "Conceptualizing the European Union as an International Actor: Narrowing the Theoretical Capability-Expectation Gap." Journal of Common Market Studies 37:3 (1999): 430; Sjursen, Helene. "What Kind of Power?" Journal of European Public Policy 13:2 (2006): 169.

[40] Bull, Hedley H. "Civilian Power Europe: A Contradiction in Terms?" Journal of Common Market Studies 21:2 (1982): 149–164; Hill, Christopher. "The Capability-Expectations Gap, or Conceptualizing Europe's External Role." Journal of Common Market Studies 31:3 (1993): 305–328.

[41] Bretherton, Charlotte and John Vogler. The European Union as a Global Actor. Second Edition. London New York: Routledge 2006: 16.

[42] Ginsberg, Roy H. "Conceptualizing the European Union as an International Actor: Narrowing the Theoretical Capability-Expectation Gap." Journal of Common Market Studies 37:3 (1999): 429–454.; White, Brian. "The European Challenge to Foreign Policy Analysis." European Journal of International Relations 5:1 (1999): 37–66; Smith, Karen E. European Union Foreign Policy in a Changing World. Second Edition. Cambridge: Polity Press 2008.

[43] Bretherton, Charlotte and John Vogler. The European Union as a Global Actor. Second Edition. London New York: Routledge 2006: 12.

[44] Smith, Michael. "Comment: Crossroads or Cul-de-sac? Reassessing European Foreign Policy." Journal of European Public Policy 13:2 (2006): 322.

[45] Duchêne, François. "Europe's Role in World Peace." European Tomorrow: Sixteen Europeans Look Ahead. R. Mayne (ed.) London: Fontana 1972: 32–47.
[46] Harnisch, Sebastian. "Außenpolitiktheorie nach dem Ost-West Konflikt: Stand und Perspektiven der Forschung." September 2002. Trierer Arbeitspapiere zur Internationalen Politik Nr. 7. 1 December 2010 *http://www.deutsche-aussenpolitik.de/resources/tazip/tazip7.pdf.*
[47] Maull, Hans W. "Germany and Japan: The New Civilian Powers." Foreign Affairs 69:5 (1990): 91–106.
[48] Manners, Ian. "Normative Power Europe: A Contradiction in Terms?" Journal of Common Market Studies 40:2 (2002): 237–241.
[49] Hyde-Price, Adrian. "Normative" Power Europe: a Realist Critique. Journal of European Public Policy 13:2 (2006): 218.
[50] Posen, Barry R. "ESDP and the Structure of World Power." The International Spectator 39:1 (2004): 5–17.
[51] Smith, Karen E. "The European Union: A Distinctive Actor in International Relations." The Brown Journal of World Affairs 9:2 (2003): 103–113; Smith, Karen E. European Union Foreign Policy in a Changing World. Second Edition Fully Revised and Updated. Cambridge: Polity Press, 2008.
[52] Sjursen, Helene. "What Kind of Power?" Journal of European Public Policy 13:2 (2006): 170.
[53] "Security" is understood here in a narrow sense of preserving the physical integrity of a country through military means. It has to be noted though that "security" can be defined in a much broader way, encompassing not only military, but also social, economic, ecological, human etc. dimensions.
[54] White, Brian. "The European Challenge to Foreign Policy Analysis." European Journal of International Relations 5:1 (1999): 45ff. It has to be noted though, that the Lisbon Treaty formerly abolished the three-pillars structure to create a single European Union. However, the factual difference between communautarised policy areas and issues dealt at the intergovernmental level is not likely to disappear in the near future.
[55] Council of the European Union. Council Decision on the Signing of the Framework Agreement Between the European Community and the European Space Agency. Doc. 12858/03 of 7 October 2003. Brussels: European Union.
[56] Britz, Malena. "Translating EU Civil Protection in the Nordic States – Towards a Theoretical Understanding of the Creation of European Crisis Management Capacities." Conference Paper. European Union Studies Association's Tenth Biennial International Conference. Montreal, Canada. 17–19 May 2007.
[57] Czempiel, Ernst-Otto. Internationale Politik. Ein Konfliktmodell. Paderborn: Schöningh 1981: 13–22.
[58] Quoted in: Smith, Karen E. European Union Foreign Policy in a Changing World. Second Edition. Cambridge: Polity Press 2008: 3.
[59] Defined by Caporaso and Jubille, quot. in: Ginsberg, Roy H. "Conceptualizing the European Union as an International Actor: Narrowing the Theoretical Capability-Expectation Gap." Journal of Common Market Studies 37:3 (1999): 447.
[60] Allen, David and Michael Smith. "Western Europe's Presence in the Contemporary International Arena." Review of International Studies 16:1 (1990): 19–37.
[61] Bretherton, Charlotte and John Vogler. The European Union as a Global Actor. Second Edition. London New York: Routledge 2006: 24ff.
[62] *Ibid.*: 24.
[63] The security dilemma is a situation where the increase of one state's security decreases the security of the other states, even if this happens inadvertently.
[64] Office of Science and Technology Policy, Executive Office of the President, The White House. U.S. National Space Policy. 31 August 2006. 6 December 2010. *http://www.globalsecurity.org/space/library/policy/national/us-space-policy_060831.pdf.*
[65] Bretherton, Charlotte and John Vogler. The European Union as a Global Actor. Second Edition. London New York: Routledge, 2006: 27.
[66] *Ibid.*: 27.

[67] *Ibid.*: 37ff.
[68] Cherkas, Sergey V. International Global Monitoring Aerospace System IGMASS – New Approach to Disaster Management Issue. Presentation to the STSC of UNCOPUOS on 17 June 2010. Vienna: UNCOPUOS. 6 December 2010. *http://www.oosa.unvienna.org/pdf/pres/copuos2010/tech-32E.pdf.*
[69] See for example: Gleason, Michael. P. "Shaping the Future with a New Space Power: Now is the Time." High Frontier 6.2. February 2010.
[70] Bretherton, Charlotte and John Vogler. The European Union as a Global Actor. Second Edition. London New York: Routledge, 2006: 29.
[71] *Ibid.*: 30.
[72] In particular Article 2 TUE laying down the EU's basic values, and Article 3.1 TUE stating that "The Union's aim is to promote peace, its values and the well-being of its peoples".
[73] Council of the European Union. Resolution on the European Space Policy. Doc. 10037/07 of 25 May 2007. Brussels: European Union.
[74] European Union. A Secure Europe in a Better World. European Security Strategy. 12 December 2003. Brussels: European Union.
[75] See the contribution of Marcus Hornung in this book.
[76] Council of the European Union. 7th Space Council Resolution. Global Challenges: Taking Full Benefit of European Space Systems. 25 November 2010. Brussels: European Union.
[77] Council of the European Union. Council Resolution. Taking Forward the European Space Policy. Doc. 13569/08 of 26 September 2008. Brussels: European Union.
[78] Commission of the European Communities. Communication from the Commission to the Council and the European Parliament. European Space Policy. COM(2007) 212 final of 26 April 2007. Brussels: European Union.
[79] *Ibid.*
[80] Johnson, Rebecca E. Europe's Space Policies and their Relevance to ESDP. 2006. Note for the European Parliament's Subcommittee on Security and Defence. 14 December 2010. *http://www.acronym.org.uk/space/PE381369EN.pdf.*
[81] Gleason, Michael P. European Union Space Initiatives: The Political Will for Increasing European Space Power. Astropolitics 4:1 (2006): 7–41.
[82] Schörnig, Niklas. „Neorealismus." Theorien der Internationalen Beziehungen. 2. Auflage. Schieder, Siegfrid and Manuela Spindler (eds.). Opladen-Farmington Hills: Verlag Barbara Budrich 2006: 65–92.
[83] Rittberger, Volker. Approaches to the Study of Foreign Policy Derived from International Relations Theories. Tübinger Arbeitspapiere zur internationalen Politik und Friedensforschung Nr. 46. Tübingen: University of Tübingen 2004.
[84] See, for example European Parliament. Report on Space and Security. Doc. 2008/2030 (INI) of 10 June 2008. Brussels: European Union; Commission of the European Communities. Commission Working Document. European Space Policy Progress Report. COM(2008)561 final of 11 September 2008. Brussels: European Union.
[85] Shaber and Ulbert quoted in160;: Boekle, Henning; Rittberger, Volker and Wolfgang Wagner. Normen und Außenpolitik: Konstruktivistische Außenpolitiktheorie. Tübinger Arbeitspapiere zur internationalen Politik und Friedensforschung Nr. 34. Tübingen: University of Tübingen 1999: 4.
[86] *Ibid.*: 7.
[87] Boekle, Henning; Rittberger, Volker and Wolfgang Wagner. Normen und Außenpolitik: Konstruktivistische Außenpolitiktheorie. Tübinger Arbeitspapiere zur internationalen Politik und Friedensforschung Nr. 34. Tübingen: University of Tübingen 1999.
[88] These are the 1969 Outer Space Treaty, the 1968 Rescue Agreement, the 1972 Liability Convention, the 1975 Registration Convention and the 1979 Moon Treaty.
[89] See, for example European Parliament. Report on Space and Security. Doc. 2008/2030 (INI) of 10 June 2008. Brussels: European Union.; Council of the European Union. Draft Code of Conduct for Outer Space Activities. Doc. 16560/08 of 3 December 2008. Brussels: European Union.

[90] Manners, Ian. "Normative Power Europe: A Contradiction in Terms?" Journal of Common Market Studies 40:2 (2002): 242.
[91] Council of the European Union. 7th Space Council Resolution: "Global Challenges: taking full benefits of European space systems". 25 November 2010. Brussels: European Union.
[92] Bretherton, Charlotte and John Vogler. The European Union as a Global Actor. Second Edition. London New York: Routledge, 2006: 60.
[93] Smith, Karen E. European Union Foreign Policy in a Changing World. Second Edition Fully Revised and Updated. Cambridge: Polity Press, 2008: 23.
[94] Commission of the European Communities. Communication from the Commission to the Council and the European Parliament. European Space Policy. COM (2007) 212 final of 26 April 2007. Brussels: European Union: 5 (Emphasis added).

Global politics

1.3 Europe as an actor in the United Nations Committee on the Peaceful Uses of Outer Space

Werner Balogh

1.3.1 Introduction

The United Nations Committee on the Peaceful Uses of Outer Space (COPUOS) is the only body in the United Nations system exclusively dealing with the peaceful uses of outer space. Several European countries are member States of the Committee, as are the European Space Agency (ESA) and a number of other European governmental and non-governmental organisations that participate as permanent observers of the Committee. On several occasions the European Union, represented by the European Commission, has also been joining the COPUOS meetings as a temporary observer.

Space activities are a particularly visible area in which Europe can shape its identity. This article will discuss the role of Europe in the Committee, in particular in the sessions following the Third United Nations Conference on the Exploration and Peaceful Uses of Outer Space (UNISPACE III), held in the summer of 1999. It will also analyse to what extent Europe, as an actor in the Committee, has contributed to building a European identity through its involvement in space activities. We will therefore mainly consider the European identity through space in its projective perspective.[95]

1.3.2 United Nations Committee on the Peaceful Uses of Outer Space

The launch of the first artificial Earth-orbiting satellite, Sputnik I, on 4 October 1957 heralded the beginning of the space age. Space activities, predicted only decades earlier by a few believing space pioneers and described by science fiction authors, had finally, and to many quite surprisingly, become a reality. Immediately, a number of important questions were raised by the international community. As the space age began in the midst of the Cold War, some expressed concerns that this may be the opening up of a new era of confrontation in outer space between the superpowers. Was there anything that could be done to prevent outer space from becoming a future battlefield? Should not there be rules and legal norms applicable to activities in outer space and if so, who would define, regulate and implement them? At the same time the great potential of space exploration and its applications to bring immense benefits to all humankind was obvious. Space-based applications such as Earth observation, navigation and telecommunications had been described before and were now in the realm of the possible. However, how could it be ensured that these benefits would accrue to all countries and not only to those that had the knowledge and the resources to actively engage in space activities?

It was under these circumstances that the international community agreed that it was within the existing mandate of the United Nations to consider the issues of international cooperation related to outer space activities. At its thirteenth session in December 1958, the General Assembly decided to take action and established the ad hoc Committee on the Peaceful Uses of Outer Space.[96] The ad hoc Committee prepared a comprehensive report on the potential of space activities and on how the United Nations and its organisations could and should be involved. Consequently, at its fourteenth session, the General Assembly decided to transform COPUOS into a permanent body and gave it the mandate to "review, as appropriate, the area of international co-operation, and to study practical and feasible means for giving effect to programmes in the peaceful uses of outer space which could appropriately be undertaken under United Nations auspices", as well as to "study the nature of legal problems which may arise from the exploration of outer space".[97]

Given the wide range of tasks before it, the Committee subsequently decided to establish two subsidiary bodies, the Legal Subcommittee and the Scientific and Technical Subcommittee, both of which held their first sessions in 1962. The Committee and its Subcommittees continue to meet based on an annual meetings schedule. The Committee is supported in its work by the United Nations Office for Outer Space Affairs of the Secretariat of the United Nations (OOSA) that was relocated from the United Nations Headquarters in New York to the United Nations Office at Vienna in 1993.[98] Consequently, the annual sessions of the Committee and its Subcommittees are also held at the United Nations Office at Vienna.

The work of COPUOS and its Subcommittees resulted in the adoption and ratification of the United Nations treaties and principles on outer space,

Fig. 1.3. *The United Nations Office for Outer Space Affairs is located at the United Nations Office at Vienna in Austria where the annual session of the Committee on the Peaceful Uses of Outer Space and its subsidiary bodies are being held. (©WTV/Popp & Hackner).*

which contribute to guide the space activities of the Member States of the United Nations and which also maintain the use of outer space for peaceful purposes.[99] It has also led to the organisation of a series of United Nations Conferences on the Exploration and the Peaceful Uses of Outer Space, the so-called UNISPACE Conferences, at which Member States have charted the way forward for space activities for the following decades. A particular notable event was the establishment in the early 1970s of the United Nations Programme on Space Applications, implemented by the Office for Outer Space Affairs, which works to assure that space benefits accrue to all humankind. As we will see, COPUOS and its subsidiary bodies continue to play an essential role as the only forum of the United Nations exclusively dealing with the issue of international cooperation in the exploration and the peaceful uses of outer space.

1.3.3 Europe and COPUOS

At its first meeting in 1958, the ad-hoc Committee consisted of 18 Member States. Over the years, the membership increased and new countries were added. In 2011, the Committee welcomed Tunisia as its 70th member.[100] Belgium, Czechoslovakia (now the Czech Republic), France, Italy, Poland, Sweden, the United Kingdom of Great Britain and Northern Ireland were among the initial COPUOS members. Other European countries joined at later sessions of the Committee. For some years, the Committee limited membership to a fixed number of seats which required some states to be represented on a rotational basis. This practice was abolished in 2001.

In addition to the permanent members of the Committee, COPUOS is also granting permanent observer status to regional and international governmental and non-governmental organisations with space-related activities. Several European governmental and non-governmental organisations have applied for and been granted permanent observer status: the European Space Agency (ESA), the European Organisation for Astronomical Research in the Southern Hemisphere (ESO), the European Association for the International Space Year (Eurisy), the European Space Policy Institute (ESPI), and the European Telecommunications Satellite Organisation (Eutelsat).

The European Union is neither a permanent member nor a permanent observer to the Committee. While the European Union, represented by the European Commissions, is in fact a permanent observer at the General Assembly, this status does not automatically extend to the Committee. However, as a courtesy, COPUOS has granted relevant entities to join its sessions as temporary observers on several occasions. The European Union has requested and been granted such observer status at several past sessions of the Committee and its subsidiary bodies.

Given the long-standing and active role of ESA as the space agency of many European countries and the role of the European Union in implementing, together with ESA and the national space agencies of the European countries, a European Space Policy, Europe is represented in the Committee at three levels: as individual member States, through ESA and through the European Union. This circumstance is however complicated by the fact that not all European countries are members of COPUOS, of ESA or of the European Union (see Table 1.1).

Decisions within the Committee are taken based on the consensus principle.[101] Even though consensus decisions are limited to the pool of permanent members of the Committee, it remains a challenge to demonstrate a strong, unified European space identity in the Committee because European countries can contribute to the discussions as individual member States, through ESA and also through statements made by the European Union. The situation is further complicated by the fact that some European Member States belong to the Western European and Others Group (WEOG) while some other belong to the Eastern European Group (EEG), two of the five unofficial Regional Groups in the United Nations that act as voting blocs and negotiation forums.[102] Several Easter European countries that are now Member States of the European Union have retained their membership in the Easter European Group. While today the roles of WEOG and EEG for consensus finding in the Committee are limited,

Table 1.1: *European countries and their membership in the European Union, in the European Space Agency (countries with a cooperation agreement with ESA are indicated with square brackets, those that have signed the European Cooperating States agreement are indicated by round brackets) and in the United Nations Committee on the Peaceful Uses of Outer Space (as of January 2011)*

Country	UNCOPUOS	ESA	European Union
Albania	1959	–	–
Austria	1959	1987	1995
Belgium	1958	1975	1951
Bulgaria	1959	–	2007
Croatia	–	–	Candidate Country
Cyprus	–	[2009]	2004
Czech Republic	1958	2008	2004
Denmark	–	1975	1973
Estonia	–	(2009)	2004
Finland	–	1995	1995
France	1958	1975	1951
Germany	1973	1975	1951
Greece	1980	2005	1981
Hungary	1959	(2003)	2004
Ireland	–	1975	1973
Italy	1958	1975	1951
Latvia	–	[2009]	2004
Lithuania	–	[2010]	2004
Luxembourg	–	2005	1951
Former Yugoslav Republic of Macedonia	–	–	Candidate Country
Malta	–	–	2004
Netherlands	1977	1975	1951
Norway	–	1986	–
Poland	1958	(2007)	2004
Portugal	1994	2000	1986
Romania	1959	(2006)	2007

Slovakia	2001	[2010]	2004
Slovenia	–	(2010)	2004
Spain	1080	1975	1986
Sweden	1958	1975	1995
Switzerland	2007	1975	–
Turkey	1977	[2004]	Candidate Country
Ukraine	2004	[2008]	–
United Kingdom	1958	1975	1973

the Regional Groupings continue to play an important role in the selection of the rotating chairperson for the Committee and its Subcommittees.

European countries as individual states obviously pursue their own national interests when it comes to bilateral international space cooperation. As members of the European Space Agency they decide upon and contribute to implement the mandatory and optional programmes and activities of ESA, which also include wide-ranging international cooperation efforts. As member of the European Union they also contribute to shaping and implementing the European Space Policy and what is to become the European Space Programme. When European countries participate in the discussion in COPUOS, their positions are therefore formed by their individual views and possibly as well by their views as ESA and/or European Union members. There are of course efforts prior to the discussions in the Committee to ensure common positions among European Member States, ESA and the European Union on issues that are considered to be of vital importance for Europe. For example, ESA organises side-meetings of its International Relations Committee (IRC) to coordinate the views of its Member and Associated States during the beginning of the first week of the annual sessions of the Committee and its Subcommittees. When attending sessions of the Committee, the European Union representatives are also participating in those meetings. More recently European member States of COPUOS that have held the European Union Presidency have also begun to make statements on behalf of the European Union in the Committee.

1.3.4 Europe's role in the Committee

Throughout the history of COPUOS, Europe has left its marks on the work of the Committee. In the 1960s and early 1970s, the discussions in COPUOS were overshadowed by the East-West confrontation. Not surprisingly, the discussions in COPUOS were initially dominated by the two space-faring superpowers. Europe's own space activities were not yet well defined and still limited to national activities. It would take Europe some time to successfully bundle its forces and to develop its own, independent space activities which allowed it to catch up with the superpowers. While Europe has not succeeded – despite attempts in the late 1980s – to develop independent human spaceflight capabilities, it has been able to develop its own launch capabilities and today has at its hands all the means and capabilities required to engage in space science and space application missions that are matching the space activities of the other spacepowers, thus making it an important potential partner for international space cooperation.

Despite this "late start" Europe started to play an important role right from the beginning of the work of the Committee with the selection of European candidates as chairpersons of the Committee and its Subcommittees. Prior to 1996, the chairmanship of the Committee and its Subcommittees was firmly assigned to specific countries, finely balancing the various political interests of Member States in the Committee. The neutral Austria provided the chairperson for the Committee, Romania, representing the Eastern Bloc, the vice-chairperson, and Brazil the rapporteur. From 1962 to 1982, the Legal Subcommittee was chaired by Poland and then by Czechoslovakia which was later replaced by the Czech Republic. Australia, with the exception of the sixth session in 1969, provided the chairperson for the Scientific and Technical Subcommittee. During the "law-making" phase of the Committee which resulted in the conclusion of five space treaties and agreements and five declarations and legal principles, the COPUOS chairmanship was therefore firmly in European hands.

The end of the Moon Race and the conclusion of the law-making phase in the Committee gave way to considerations influenced by the increased self-confidence of countries that had gained independence. In addition, an emerging North-South divide in the late 1970s also had a strong influence on the direction of the discussions in COPUOS.

The end of the East-West confrontation with the dissolution in the late 1980s of the Eastern Bloc, the former communist states of Eastern and Central Europe (the Soviet Union and its satellites in the Warsaw Pact), not only marked the beginning of a new geopolitical situation, but also the start for taking the first steps towards a European Union freed from the mentality of the Cold War bloc thinking.

The practice of assigning chairperson positions to particular member States in the Committee was discontinued in 1996 and replaced by a system of rotation which aims to ensure the equitable representation of the various geographical regions. This change also reflected the newly existing international political realities.

1.3.4.1 Europe and the UNISPACE conferences and their follow-ups

The special situation of Austria as a neutral country during the Cold War and its chairmanship of COPUOS until 1996, as well as the country's role as the host country for the United Nations Office at Vienna likely had an influence on the decisions to hold the three UNISPACE conferences in Vienna. These conferences played an important role in deliberating and setting the agenda of the international community for international space cooperation.

The detailed recommendations of these conferences are contained in their voluminous reports and the implementation of the recommendations can be re-traced through the annual reports of COPUOS and its Subcommittees. Only some of the major outcomes will therefore be mentioned in the following.

For example, the recommendations of the first UNISPACE conference held in 1968 led to the creation of the United Nations Programme on Space Applications and to the establishment of the post of the United Nations Expert on Space Applications within the Office for Outer Space Affairs. Through these measures, the United Nations was enabled to provide support to those countries that lacked the human, financial and technical resources necessary to fully utilise the benefits of space technology.[103,104] To date approximately 15.000 space professionals have participated in the activities of the Programme and 300 specialists have participated in various long-term fellowship programmes. The Programme continues to organise an annual series of seminars, symposiums, training courses and workshops.[105]

The second UNISPACE Conference held in 1982 (UNISPACE'82) broadened the mandate of the Programme on Space Applications[106] and led to the establishment of four Regional Centres for Space Science and Technology Education, affiliated to the United Nations.[107,108] The Centres are located in the regions that correspond to the United Nations Economic Commissions for Africa, Asia and the Pacific, and Latin America and the Caribbean. The plan to establish a similar centre in Eastern Europe was eventually abandoned following the dissolution of the Eastern Bloc. Discussions on establishing a centre in the Middle East are continuing.

The third and latest UNISPACE Conference (UNISPACE III) was held in 1999 and concluded with "The Space Millennium: Vienna Declaration on Space

Table 1.2: *Chairpersons from European member States of COPUOS in COPUOS, Scientific and Technical Subcommittee, Legal Subcommittee and Working Groups in the period following the UNISPACE III Conference (2000–2010)*

Entity	Year(s)	Chairperson(s)	Relevant Reports
Committee on the Peaceful Uses of Outer Space	2005–2006 2010–2011	Gerard Brachet (France) Dimitru-Dorin Prunariu (Romania)	A/60/20; A/60/20/ Corr.1; A/61/20 A/65/20
Legal Subcommittee	2000–2003 2008–2009 2004–2005	Vladimir Kopal (Czech Republic) Sergio Marchisio (Italy)	A/AC.105/738; A/AC.105/763; A/AC.105/787; A/AC.105/805; A/AC.105/917; A/AC.105/935 A/AC.105/826; A/AC.105/850
Working Group on Status and Application of the Five United Nations Treaties on Outer Space	2002–2004 2006–2009 2010	Vassilios Cassapoglou (Greece) Jean François Mayence (Belgium)	A/AC.105/787, Annex I; A/AC.105/805, Annex I; A/AC.105/826, Annex I; A/AC.105/871, Annex I; A/AC.105/891, Annex I; A/AC.105/917, Annex I; A/AC.105/935, Annex I; A/AC.105/942, Annex II
Working Group on Review of the Concept of the 'Launching State'	2000–2002	Kai-Uwe Schrogl (Germany)	A/AC.105/738, Annex II; A/AC.105/763, Annex II; A/AC.105/787, Annex IV; A/AC.105/826, Appendix II; A/RES/59/115
Working Group on Examination of the Preliminary Draft Protocol on Matters Specific to Space Assets to the Convention on International Interests in Mobile Equipment, Opened for Signature at Cape Town, South Africa, on 16 November 2001	2003 2004–2005	Sergio Marchisio (Italy) Vladimír Kopal (Czech Republic)	A/AC.105/805, Annex III; A/AC.105/826, Annex III; A/AC.105/850, Annex II
Working Group on Practice of States and International Organizations in Registering Space Objects	2005 2006–2007	Niklas Hedman (Sweden) Kai-Uwe Schrogl (Germany)	A/AC.105/850, Annex III; A/AC.105/871, Annex III; A/AC.105/891, Annex III; A/RES/62/101

Working Group on General Exchange of Information on National Legislation Relevant to the Peaceful Exploration and Use of Outer Space	2008–2011	Irmgard Marboe (Austria)	A/AC.105/935, Annex III; A/AC.105/942, Annex III
Scientific and Technical Subcommittee	1996–2000 2004–2005 2010–2011	Dietrich Rex (Germany) Dimitru-Dorin Prunariu (Romania) Ulrich Huth (Germany)	A/AC.105/736; A/AC.105/823; A/AC.105/848; A/AC.105/958
Working Group on Nuclear Power Sources	2000–2011	Sam A. Harbison (United Kingdom of Great Britain and Northern Ireland)	A/AC.105/736, Annex III; A/AC.105/761, Annex III; A/AC.105/786, Annex III; A/AC.105/804, Annex IV; A/AC.105/823, Annex IV; A/AC.105/848, Annex III; A/AC.105/869, Annex III; A/AC.105/890, Annex II; A/AC.105/911, Annex II; A/AC.105/933, Annex II; A/AC.105/934; A/AC.105/958, Annex II
Working Group on Near-Earth Objects	2007 2008–2009	Richard Tremayne-Smith (United Kingdom of Great Britain and Northern Ireland) Richard Crowther (United Kingdom of Great Britain and Northern Ireland)	A/AC.105/890, Annex III A/AC.105/911, Annex III; A/AC.105/933, Annex III
Working Group on Space Debris	2004–2007	Claudio Portelli (Italy) Petr Lála (Czech Republic) (2004)	A/AC.105/823, Annex III; A/AC.105/848, Annex II; A/AC.105/869, Annex II; A/AC.105/890, Annex IV

and Human Development", adopted by the Conference and endorsed by the General Assembly.[109–111] Its implementation resulted in several achievements, including the International Charter "Space and Major Disasters", the establishment of the International Committee on Global Navigation Satellite Systems (ICG) and the establishment of the United Nations Platform for Space-based Information for Disaster Management and Emergency Response (UN-SPIDER).[112–114]

Europe is playing an important role as a major contributor to the implementation of these activities. European countries frequently host activities of the Space Applications Programme. Since many years the European Space Agency is the largest single general sponsor of the United Nations Programme on Space Application, co-organising and co-funding several of its annual activities. For some time ESA also offered a series of long-term fellowship programmes to qualified candidates from developing countries. Occasionally the European Commission, in the late 1990s, has provided co-funding to activities of the Programme. While this is not the case currently, some of the activities of the Programme have thematic links to space-related activities funded by the European Union, including the planned Galileo Global Navigation Satellite System, the Global Monitoring for Environment and Security (GMES) initiative as well as a number of projects funded through the European Union's Framework Programmes.

The International Charter "Space and Major Disasters" was initiated by ESA and the French Space Agency CNES. Between 2007 and 2009, 33% of the activations were made through the United Nations. With Galileo and the European Geostationary Navigation Overlay Service (EGNOS) Europe is represented in the Providers Forum of the ICG by the European Union. UN-SPIDER receives the major part of its funds from Austria and Germany, with one of the two non-Vienna based UN-SPIDER offices being located at the United Nations campus in Bonn.

European Member States and the European Space Agency (after its founding in 1975) have contributed to the implementation of the recommendations of the UNISPACE conferences. The specific question of how this may contribute to building a European identity obviously only becomes relevant for the time following the UNISPACE III conference.

During the discussions on how to implement the recommendations of UNISPACE III, the Committee agreed to revise the structure of the agendas of the Scientific and Technical Subcommittee and the Legal Subcommittee. It introduced the possibility of multi-year items under workplans and single issue items to be considered for one session. As we will see, many of the Working Groups established under the multi-year items were and are being chaired by European chairpersons, manifesting the role of Europe as an actor in the Committee (see Table 1.2). The following section will be dedicated to those agenda items and their Working Groups.

1.3.4.2 Europe's role in the Scientific and Technical Subcommittee

The use of Nuclear Power Sources (NPS) in outer space raises many safety-related questions. The Working Group on the Use of Nuclear Power under the chairmanship of Sam A. Harbison (United Kingdom of Great Britain and Northern Ireland) has been discussing NPS-related issues for many years. In 2009, the Joint Expert Group of the Scientific and Technical Subcommittee and the International Atomic Energy Agency (IAEA), established by the Working Group, concluded their work on a "Safety Framework for Nuclear Power Sources Applications in Outer Space" which was adopted by the Subcommittee, endorsed by the Committee and agreed to by the IAEA Commission on Safety Standards. That Safety Framework was subsequently jointly published by the Subcommittee and the International Atomic Energy Agency.[115] The contribution of the European Space Agency to the development of the Safety Framework was specifically cited by the Working Group.[116] The Working Group is now focussing on discussing ways of promoting and facilitating the implementation of the Safety Framework.

A major issue of concern to the future of space activities is the threat posed by space debris. Many years of discussions on this topic in the Committee eventually resulted in an agreement on the "Space Debris Mitigation Guidelines of the Committee on the Peaceful uses of Outer Space".[117] The European Space Agency and several of the national European space agencies contributed to the space debris discussions, either directly or through their participation in the Inter-Agency Space Debris Coordination Committee (IADC). The Working Group on Space Debris, which had developed the space debris mitigation guidelines adopted by the Committee, was chaired by Claudio Portelli (Italy).[118]

To address the issue of Near-Earth Objects, the Scientific and Technical Subcommittee established a one-year Working Group on Near-Earth Objects (NEOs) in 2007, under the chairmanship of Richard Tremayne-Smith (United Kingdom of Great Britain and Northern Ireland). The Working Group subsequently prepared a three-year workplan with the aim to draft and agree on international procedures for NEO threat handling. In 2008 and 2009 that Working Group was chaired by Richard Crowther (United Kingdom of Great Britain and Northern Ireland). In 2010 the Working Group presented draft recommendations for an international response to the threat of NEO impact.[119] Discussion on these recommendations will continue at the 2011 session of the Scientific and Technical Subcommittee.

A recent new multi-year item under a workplan is focusing on "Long-term sustainability of outer space activities". It was proposed by France and supported by the other European member States in the Committee.[120] Under this item, the

Working Group will discuss potential measures to enhance the long-term sustainability of space activities and aims to prepare a document entitled "Best practices for the long-term sustainability of outer space activities", to be presented to the Committee in 2012. While not related, the French proposal, based on the initiative of the former COPUOS chairperson Gérard Brachet (France), to have this item included on the agenda of the Scientific and Technical Subcommittee, was for some time also brought in connection with the initiative of the Council of the European Union to prepare a draft Code of Conduct for outer space activities, which was discussed during the French EU Presidency in the second half of 2008.[121]

1.3.4.3 Europe's role in the Legal Subcommittee

Discussions on the "Status and Application of the Five United Nations Treaties on Outer Space" have been taking place under a long-standing agenda item in the Legal Subcommittee under the chairmanship of Vassilios Cassapoglou (Greece). In a working paper submitted in 1998 by Germany on behalf of Austria, Belgium, the Czech Republic, Denmark, Finland, France, Greece, Hungary, Ireland, Italy, the Netherlands, Norway, Poland, Portugal, Romania, Spain, Sweden, Switzerland and the United Kingdom of Great Britain and Northern Ireland, those states informed the Committee about the status of their ratification of the space treaties and expressed their views on the issues to be discussed under this agenda item.[122] Among the suggestions made, one called for improvements to the Registration Convention.

On the basis of General Assembly Resolution 1721 B (XVI) on "International co-operation in the peaceful uses of outer space", of 1961, and on the basis of the Registration Convention of 1976, launching states are called upon to register their objects launched into outer space with the United Nations. The increased number of actors participating in launch activities, including from the private sector, led in some cases to the non-registration of space objects. As a consequence, some member States in the Committee felt that there was a need to review the concept of the "launching state" as well as registration procedures. Following a proposal submitted by Germany on behalf of Austria, Belgium, the Czech Republic, Denmark, Finland, France, Greece, Hungary, Ireland, Italy, the Netherlands, Norway, Poland, Portugal, Romania, Spain, Sweden, Switzerland and the United Kingdom of Great Britain and Northern Ireland and following intersessional consultations in Bonn on 9 December 1998, the Legal Subcommittee agreed to conduct a review of the concept of the launching state under a three-year work plan, during its sessions from 2000 to 2002.[123,124] At its session in 2000 the Committee established a Working Group on "Review of the concept of the 'launching state'" under the chairmanship of Kai-Uwe Schrogl (Germany). The work of the Working Group resulted in General Assembly Resolution 59/115 on the Application of the concept of the "launching state" of 10 December 2004, clarifying the term "launching state" in view of new developments and new space actors.

Similarly, on the basis of a working paper submitted by Australia, Austria, Canada, the Czech Republic, France, Germany, Greece, India, Japan, the Netherlands, Sweden, Ukraine, the United Kingdom of Great Britain and Northern Ireland and the United States of America, the Legal Subcommittee in 2003 agreed to consider a new agenda item entitled "Practice of States and international organizations in registering space objects" under a 4-year workplan with the aim to draft recommendations for enhancing adherence to the Registration Convention.[125] In accordance with the workplan, the Legal Subcommittee in 2005 established a Working Group under this agenda item under the chairmanship of Niklas Hedman (Sweden) and under the chairmanship of Kai-Uwe Schrogl (Germany) from 2006 to 2007. The work of the Working Group resulted in General Assembly Resolution 62/101 on the "Recommendations on enhancing the practice of States and international intergovernmental organisations in registering space objects" of 17 December 2007, which aims to enhance the utility of the Register, the adherence to the Registration Convention, the acceptance also by international organisations, and to harmonise registration practices (e.g. designators, units, UTC) and encourage launching states to provide more detailed information as well as information about later changes in the supervision of space objects.

At the session of the Subcommittee in 2007, some member States expressed the view that under the agenda item entitled "Status and Application of the Five United Nations Treaties on Outer Space" consideration should be given to the reasons for the low participation of states in the fifth space treaty, the "Agreement Governing the Activities of States on the Moon and Other Celestial Bodies" (the "Moon Agreement") and that efforts should be made to resolve any identified obstacles to their participation.[126] At the session in 2008, Austria, Belgium, Chile, Mexico, the Netherlands, Pakistan and the Philippines, all of them states parties to the Agreement, prepared a joint statement on the benefits of adherence to the Moon Agreement based on an initiative of Belgium.[127] In their statement these States parties to the Moon Agreement encourage states to sign and ratify the Agreement, in particular

Fig. 1.4. *Working Group on National Legislation Relevant to the Peaceful Exploration and Use of Outer Space under the Chairmanship of Prof. Irmgard Marboe (Austria) at the 48th session of the Legal Subcommittee of the Committee on the Peaceful Uses of Outer Space, held in Vienna from 23 March to 3 April 2009 (© Rafael Aquilar Moro).*

considering their possible involvement in future missions or projects aimed at exploring celestial bodies.

For the period 2008–2011 the Legal Subcommittee is considering an agenda item entitled "General exchange of information on national legislation relevant to the peaceful exploration and use of outer space".[128] The Working Group established under this item is chaired by Irmgard Marboe (Austria) and considers the various approaches taken by member States to transpose and implement the obligations of the space Treaties into national legislation.

It should finally also be mentioned that the Working Group established under the agenda item entitled "Examination of the preliminary draft protocol on matters specific to space assets to the Convention on International Interests in Mobile Equipment (opened for signature in Cape Town on 16 November 2001)" was chaired by Sergio Marchisio (Italy) in 2003, and by Vladimír Kopal (Czech Republic) in 2004 and 2005. Discussions on this protocol continue within the International Institute for the Unification of Private Law (UNIDROIT).

1.3.4.4 European Space Policy, Treaty of Lisbon and COPUOS

On 22 May 2007 the Council of the European Union adopted a Resolution on the European Space Policy, following its endorsement by the Fourth "Space Council" – a concomitant meeting of the Council of the European Union and of the Council of the European Space Agency.[129] The Resolution makes reference to the United Nations Outer Space Treaty and specifically to the ongoing efforts of the United Nations' Committee on the Peaceful Uses of Outer Space (COPUOS) on the mitigation and prevention of space debris. With regards to international cooperation, the Resolution "invites the European Commission, the ESA Director General and the Member States to develop and pursue a joint strategy and establish a coordination mechanism on international relations. This strategy should be consistent with Member State activities and is aimed at strengthening Europe's role in the global space field and at benefiting from international cooperation . . . " The key issues to be considered in the development of a strategy for international relations are listed in Annex III to the Resolution and include improving access to third markets for European space products and services, reducing the cost of acquiring space systems through international cooperation, enabling Europe to participate in ambitious programmes that it could not accomplish independently, attracting international partners to European conceived programmes such as Galileo and GMES and making full use of the potential of space systems for sustainable development and in support of developing countries, in particular in Africa.

The political dimension of space in Europe was further strengthened with the entry into force of the Lisbon Treaty on 1 December 2009, which gives the European Union a shared competence in space.[130] However, the Treaty in its Article 189 explicitly excludes efforts towards any harmonisation of the laws and regulations of the Member States. With the ratification of the Lisbon Treaty, the European Union has replaced and succeeded the European Community. The European Union is now exercising all rights and assuming all obligations of the European Community, including its status within the United Nations, as well as all the European Union obligations. The European Commission Delegation to the United Nations has become the European Union Delegation, under the authority of the High Representative for Foreign Affairs and Security Policy.[131] Formally the European Union Delegation to the United Nations in Vienna is responsible for matters related to COPUOS.[132]

The definition and implementation of a European Space Policy is an ongoing effort between the Member States of the European Union, the European Space Agency and the European Commission. It remains to be seen if and how it will eventually affect the activities of European countries in

COPUOS and thus contribute to promoting a European identity within the Committee.

1.3.5 Conclusions

As has been demonstrated, European member States of the Committee have been and are making active and important contributions to the work of COPUOS and its Subcommittees. While there are efforts to coordinate their activities, they are still acting as independent member States in the Committee. The presence of the European Space Agency as a permanent observer in the Committee and its activities and statements have unquestionably contributed to shape – in the eyes of non-European member States – a certain European identity with regards to European space activities. The contributions European member States and ESA have made to the implementation of the recommendations of UNISPACE III and to supporting the United Nations Programme on Space Applications, reflected in the annual COPUOS reports, have also contributed to establish Europe as an important space actor.

How the rise of the European Union as a new player in the European Space Policy will change the way in which European COPUOS member States will interact with the Committee remains to be seen. It will be up to the European Union Member States to embrace this as an opportunity to further strengthen the European identity within COPUOS. While for the present time it is not realistic to expect that the European Union will eventually replace and represent European Member States in the Committee, it is also unlikely that the European Space Policy will have no practical consequences for the European contributions to the discussions in COPUOS. The establishment of a common European Space Policy will be challenging as it touches upon areas of foreign and security policy, in which European Member States still prefer to retain their national sovereignty. Whatever that outcome will be, Europe's space involvement, as it has demonstrated on many occasions in the past, is too large for it to be ignored by the other space actors.

With respect to a participative perspective of a European identity in space, there is certainly plenty of room for considering ways on how to inform the general European public about the role of Europe in the work of COPUOS. Popular topics such as the exploration of space, the space debris issue and discussions on Near-Earth Objects do find resonance with the general public. The European public may be interested to learn and possibly surprised to hear to what extent Europe is actively involved in discussing and contributing to such issues.

It is for Europe to decide what it wants to be: a collective of countries trying to secure the status quo, to address the issues of its aging population and to prevent itself from falling back into darker times or a bearer of the torch paving the way for the whole world into a brighter future. In this perspective, objectives could be to pursue worthy goals that will challenge and involve our whole society, and where values such as pursuing technological progress, curiosity, the search for wisdom, and the drive to explore and innovate will be put at the forefront. This approach can bring benefits to all humankind for a better life on Earth and eventually also in outer space, because a sustainable future will require to embrace outer space. Europe's future can only be based on the creation of real knowledge and real values – today people are expecting the European decision makers to take the necessary actions.

The concept of a European identity in space is not easy to grasp and lacks a common understanding. It may therefore mean different things to different people. The author of this article, a European, recalls a particular moment during the forty-seventh session of the Legal Subcommittee, when on 3 April 2008 the Office for Outer Space Affairs in cooperation with ESA arranged for a live-transmission from the International Space Station (ISS) of the docking of ESA's Jules Verne Automated Transfer Vehicle, the first docking of an independently developed European spacecraft, with the ISS. Up to the moment when the spacecraft slowly approached the station and the commentator finally announced the successful docking the atmosphere in the conference room was filled with anxious but exciting tension. The successful docking mirrored the achievements of European technology development and the work of decades involving thousands of Europeans. In this moment it felt like the delegates, from their seats of a comfortable conference room here on Earth, had just been looking into space, seeing a preview of what our future in space may hold for us. It was a moment when the meaning and significance of a European identity in space could be vividly felt.

Note: United Nations documents quoted in this paper are available from the website of the Office for Outer Space Affairs at www.unoosa.org and from the Official Document System of the United Nations at documents.un.org.

Disclaimer: The views expressed herein are those of the author and do not reflect the views of the United Nations.

---

95 Hornung, Marcus. "European Identity Through Space – How to Make Public Opinion Instrumental." ESPI Perspectives 37, August 2010.

96 General Assembly Resolution 1348 (XIII) "Question of the Peaceful Use of Outer Space." 13 December 1958.

97 General Assembly Resolution 1472 (XIV) "International Co-operation in the Peaceful Uses of Outer Space" 12 December 1959.

98 See UNOOSA website. *http://www.unoosa.org/*.

[99] United Nations Treaties and Principles on Outer Space: Text of treaties and principles governing the activities of States in the exploration and use of outer space and related resolutions adopted by the General Assembly, ST/SPACE/11/Rev. 2, March 2008. The "Treaty on Principles Governing the Activities of States in the Exploration and Use of Outer Space, Including the Moon and Other Celestial Bodies" ('Outer Space Treaty': General Assembly Resolution 2222 (XXI), Annex); the "Agreement on the Rescue of Astronauts, the Return of Astronauts and the Return of Objects Launched into Outer Space" ('Rescue Agreement': General Assembly Resolution 2354 (XXII), Annex); the "Convention on International Liability for Damage Caused by Space Objects" ('Liability Convention': General Assembly Resolution 2777 (XXVI), Annex); the "Convention on Registration of Objects Launched into Outer Space" ('Registration Convention': General Assembly Resolution 3235 (XXIX), Annex); and the "Agreement Governing the Activities of States on the Moon and Other Celestial Bodies" ('Moon Agreement': General Assembly Resolution 34/68, Annex). The five declarations and legal principles are the "Declaration of Legal Principles Governing the Activities of States in the Exploration and Use of Outer Space" (General Assembly Resolution 1962 (XVIII)); the "Principles Governing the Use by States of Artificial Earth Satellites for International Direct Television Broadcasting" (General Assembly Resolution 37/92, Annex); the "Principles Relating to Remote Sensing of the Earth from Outer Space" (General Assembly Resolution 41/65, Annex); the "Principles Relevant to the Use of Nuclear Power Sources in Outer Space" (General Assembly Resolution 47/68); and the "Declaration on International Cooperation in the Exploration and Use of Outer Space for the Benefit and in the Interest of All States, Taking into Particular Account the Needs of Developing Countries" (General Assembly Resolution 51/122, Annex).

[100] "United Nations Committee on the Peaceful Uses of Outer Space: Members." UNOOSA website. 25 January 2011. *http://www.unoosa.org/oosa/en/COPUOS/members.html.*

[101] In the "Report of the Committee on the Peaceful Uses of Outer Space" (General Assembly Document A/5181 of 27 September 1962, Paragraph 4), it is stated: "it has been agreed among the members of the Committee that it will be the aim of all members of the Committee and its sub-committees to conduct the Committee's work in such a way that the Committee will be able to reach agreement in its work without need for voting".

[102] "Members of the General Assembly." United Nations website. 25 January 2011. *http://www.un.int/wcm/webdav/site/gmun/shared/documents/GA_regionalgrps_Web.pdf.*

[103] United Nations Office for Outer Space Affairs, ST/SPACE/52, "United Nations Programme on Space Applications", February 2010.

[104] General Assembly Resolution 2601 A (XXIV), "International co-operation in the peaceful uses of outer space", 16 December 1969.

[105] "United Nations Programme on Space Applications Activities Schedule: 2011." UNOOSA website. 25 January 2011. *http://www.unoosa.org/oosa/en/SAP/sched/index.html.*

[106] General Assembly Resolution 37/90, "Second United Nations Conference on the Exploration and Peaceful Uses of Outer Space", Para. 7: Redirection of the Programme on Space Applications, Regional Centres, 10 December 1982.

[107] General Assembly Resolution 45/72, "International Co-operation in the Peaceful Uses of Outer Space", 11 December 1990.

[108] United Nations Office for Outer Space Affairs, ST/SPACE/41, "Capacity-building in space science and technology – Regional Centres for Space Science and Technology Education, affiliated to the United Nations", December 2008.

[109] UNISPACE III website. 25 January 2011. *http://www.un.org/events/unispace3/.*

[110] A/CONF. 184/6

[111] General Assembly Resolution 54/68, "Third United Nations Conference on the Exploration and Peaceful Uses of Outer Space", 6 December 1999.

[112] International Charter Space and Major Disasters website. 25 January 2011. *http://www.disasterscharter.org.*

[113] International Committee on Global Navigation Satellite Systems website. 25 January 2011. *http://www.icgsecretariat.org.*

[114] UN Spider website. 25 January 2011. *http://www.un-spider.org*.
[115] A/AC.105/934.
[116] A/AC.105/958, Annex II, para. 12.
[117] A/AC.105/890, Annex IV.
[118] A/AC.105/890, para. 101.
[119] A/AC.105/C.1/L.301, annex.
[120] A/AC.105/C.1/L.303, Long-term sustainability of outer space activities, Working paper submitted by France.
[121] Council of the European Union. Council Conclusions and Draft Code of Conduct for Outer Space Activities. Doc. 16560/08 of 3 December 2008. Brussels: European Union.
[122] A/AC.105/C.2/L.211/REV.1.
[123] A/AC.105/C.2/L.211/Rev.1.
[124] A/AC.105/L.217.
[125] A/AC.105/C.2/L.241 and Add.1.
[126] A/AC.105/891.
[127] A/AC.105/C.2/L.272, annex.
[128] General Assembly resolution 63/90.
[129] Council of the European Union. Outcome of Proceedings of the Council (Competitiveness) on 21–22 May 2007 – Resolution on the European Space Policy. Doc. 10037/07 of 25 May 2007. Brussels: European Union.
[130] Official Journal of the European Union. Consolidated Versions of the Treaty on European Union and the Treaty on the Functioning of the European Union. Doc. 2010/C 83/01 of 30 March 2010. Brussels: European Union.
[131] "European Union @ United Nations" website. 25 January 2011. *http://www.europa-eu-un.org/home/index_en.htm*.
[132] "About the EU at the UN – European Union Delegations. Vienna." European Union @ United Nations. 25 January 2011. *http://www.europa-eu-un.org/articles/articleslist_s38_en.htm*.

International cooperation

1.4 Motivational and cultural conditions for regional space collaboration: is Europe an irreproducible model?

Ariane Cornell

1.4.1 Introduction

The world's space sector today sits at an important turning point. Space technology and its terrestrial applications have matured to a point where an increasing number of countries from various regions are choosing to enter into the sector. Once an extended battleground for the U.S. and the Soviets, space is now an international arena for supporting human security and economic development. According to the Space Foundation, "by the end of 2009, at least 49 countries had deployed civil satellites, either on their own or in cooperation with other nations, and more than 115 countries owned at least a share in an operating satellite. During the next five years, more than 100 missions into space are planned by civilian space agencies from more than 15 nations."[133] As space technology develops further, these numbers will continue to grow.

Most of these new countries joining the space sector come from emerging space regions such as Asia Pacific, Latin America and Africa. These countries are looking to bridge the space technology gap as fast as possible between developed space countries and themselves. Recently during political discussions among the developing space regions in forums such as the Space Conference of the Americas and the Asia-Pacific Regional Space Agency Forum (APRSAF), the idea of creating large-scale regional collaborations has emerged as a way to bring whole developing space regions up faster collectively.

The functioning example of large-scale regional collaboration to which these emerging space regions look up is that of Europe. The European space collaboration has kept Europe on a similar technological level with the Russians and the United States. Furthermore, it has contributed to strengthening the region's identity as an intellectually developed community working together for the good of its citizens. Space has particularly played into strengthening this European identity, as viewed externally and internally, because Europe's approach has been in such contrast to its Russian and American peers, who acted primarily unilaterally for most of the development of their space programmes.

Indeed, it is understandable how the European collaborative space system is one that a developing region may aspire to. This system, though, is very hard to reproduce. The European large-scale collaborative space model is unique and has worked because both the motivational and cultural conditions were correct. Both of these important sets of conditions for the moment are not present in the developing space regions of Asia Pacific, Latin America and Asia. Therefore, using Europe as a model to implement now to collectively improve their respective regions is not the right strategy.

1.4.2 Regional space collaboration – benefits and challenges

Large-scale regional space collaboration like that in Europe includes working together on many levels. It means working together in a practical sense. For Europe, this has meant cooperating on complex, multilateral technical projects such as the Ariane rockets, the Herschel space telescope, and the Galileo global navigation satellite system. The collaboration also involves aligning organisationally. In Europe, most obviously this has meant the creation of the European Space Agency (ESA). Finally, regional space collaboration also includes working together politically. In Europe, this has entailed coming together to agree on specific space questions such as, whether to support continuing the International Space Station. This has also entailed addressing broader policy questions such as technology transfer, immigration, and interregional labor laws which directly affect the operations of regional space collaborations via well established, regional political collaborations such as the European Union.

Regional collaboration systems of this scale have their benefits, which Europe has seen. First, as stated in its mandate, ESA has "coordinat[ed] the financial and intellectual resources of its members" in order to "undertake programmes and activities far beyond the scope of any single European country."[134] It is clear that without this cooperation, few to no European countries would have been able to compete with the resources of the U.S. and Russia. Second, when clearly working on the same team, it allows countries to have positive technology and knowledge exchange while trying to create efficient research systems. Smaller countries with fewer resources can still contribute by allocating their limited resources to becoming specialists in certain science or technology areas. This specialised technology can then become an asset to contribute to the collaborative system and motivates stronger countries to support the smaller countries' technology development. Finally, these collaborations are politically helpful. Productive policy decisions, whether regarding decisions on limiting space debris or rules on space situational awareness, are made more easily and quickly at the regional level. While multilateral decisions are never easy, intra-regional similarities (e.g., cultural, historical, economic, political, and linguistic) as well as simply having fewer countries around the table enable simpler negotiations than those at the global scale. Moreover, solidarity, when created on a regional level, can translate into helpful political critical mass at the international level, such as at the United Nations Committee on the Peaceful Uses of Outer Space (UN COPUOS).

Despite these benefits there are many challenges to collaborative space programmes whether they are domestic, regional or international. These issues were studied extensively in 2009 and 2010 by the American National Research Council's (NRC) Committee on the Assessment of Impediments to Interagency Cooperation on Space and Earth Science Missions. One main issue is that collaborative international space programmes, though they may be able to pool in more resources, are not as practically efficient with the resources as theory would assume. Collaborations require excellent systems management to control contingent elements of programmes. The political, financial and technical elements of programmes are more complicated in international collaborative scenarios which also lead to increased delays and risk.[135]

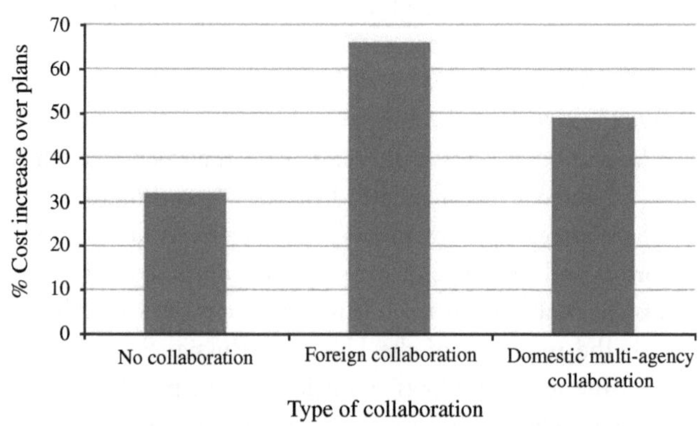

Fig. 1.5. *Space programme cost increases during collaboration versus no collaboration.*[136]

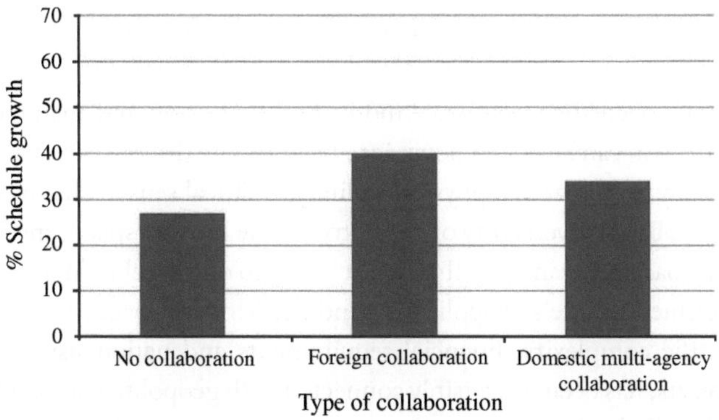

Fig. 1.6. *Space programme schedule growth (delay) during collaboration versus no collaboration.*[137]

The other challenge to multi-national collaborative regional space systems is that it is difficult to coordinate asymmetric national space programmes. Very often, the space asymmetries are indicative of differences in general technological development capabilities. These asymmetries tend to disappear particularly in large, international, political settings where countries look to other countries in their region for a sense of solidarity and community. The discrepancies become very clear, though, once the countries return to their region. This political form of relativity can make it very hard for countries with very different levels of space strength to collaborate. In regions where there are some strong players and some weak players, there must be the right incentives for all countries, particularly the strong countries, to collaborate with the weaker countries.

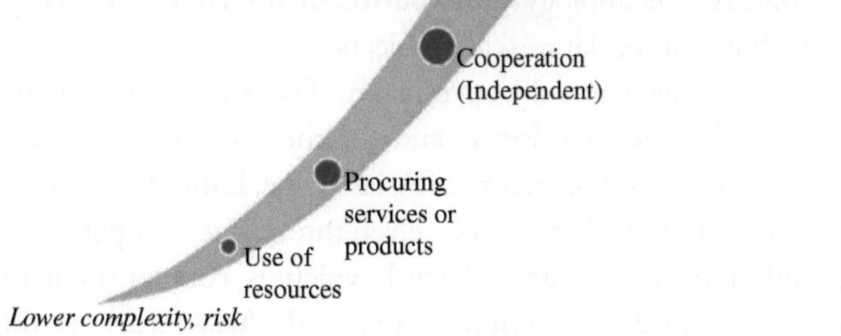

Fig. 1.7. *The relationship between increasing interdependency between multiagency participants in a collaborative mission and mission complexity and performance risks.*[138]

1.4.3 Motivational conditions for collaboration

The conditions needed to politically motivate the creation and maintenance of space collaboration can be boiled down into two primary drivers: threat and needs. The former may be threat in a physical military, political sense. That is, threat to defences and political sovereignty of a country. In the broader space sense that also includes civil space programmes, this threat is also to one's technological standing which also influences one's geopolitical standing. Modern space science, because of the scientific complexity, financial requirements and nationalism connected with the projects, has been irreversibly connected with geopolitics since it began in the Cold War. Today it continues. Countries will connect with each other when they perceive that they will have more strength en masse. This applies both to smaller countries and to countries which may be regionally strong but on a global scale are mediocre.

The second, and not independent, driver for collaboration is need. One tends to not engage in exchanging goods or services unless there is a need for the other party's offerings. This is also the case in space. As put clearly by the NRC Report, "A prerequisite for a successful international collaboration is that all parties believe the collaboration is of mutual benefit."[139] Does Country B have a technology or skill that could complement or enhance Country A's competencies? If Country B does not have it now, do they have the potential and motivation to have it in the near future? If the answer to these questions is "no," then collaboration is not interesting. According to a joint NRC-European Space Science Committee-European Science Foundation report analysing U.S.-Europe space collaboration, "only those international cooperative efforts should be attempted in which participants consider themselves partners [. . .] and have confidence in one another's reliability and competence as well as their dedication to the overall mission goals."[140] Furthermore, in the space sector, the larger the gap is between the technical competencies of countries, the less likely there is to be the sense of need and therefore collaboration. This is particularly true because the value of the other country's technology must outweigh the added efficiency burden inherent in collaboration on large-scale projects.

Politics can be a somewhat subjective realm. These motivational conditions, though, are relatively objective drivers and for the moment they do not exist properly in the developing space regions of Asia Pacific, Latin America and Africa. In Asia Pacific, it is true that there is a perceived threat, but it is operating less on a global scale and more on a regional level which is counterproductive if an intraregional collaborative programme is the goal. Asia Pacific is the fastest growing space region of the developing regions. China has sprinted up the geopolitical ladder in the past 15 years. Space has been a cornerstone in demon-

Motivational Drivers	Definition	Consequences
Threats	• Physical military, political sense: threat to defences and political sovereignty of a country • Technological sense: threat to one's technological standing which also influences one's geopolitical standing	• Countries will connect with each other when they perceive that they will have more strength en masse • This applies both to smaller countries and to countries which may be regionally strong but on a global scale are mediocre
Needs	• Need for another country's current or future technology or skill that could complement or enhance one's own competencies	• The larger the gap is between the technical competencies of countries, the less likely there is to be the sense of need • The value of the other country's technology must outweigh the added efficiency burden inherent in collaboration on large-scale projects

Fig. 1.8. *Definition of motivational drivers in the sense of regional space collaboration.*

strating financial and technical wealth as well as inspiring domestic nationalism. The same is true for another regional power, India, who has heavily invested in their space programme in the past 15–20 years, as well. Just in the past two years, the Indian Space Research Organisation (ISRO) has grown its budget more than 35% from 4.167 crore rupees (675 million euros) to 5.778 crore rupees (935 million euros) in 2010.[141] These countries are faring well and are respected in the global space sector. Smaller countries like Japan and South Korea have also done well independently. Japan was the first Asian country to join the development of the International Space Station and has been a founder of space development in the region. South Korea has a space programme that has grown in leaps and bounds for a country that was only declared a developed nation in 2008 by the FTSE global financial indexes.[142]

In Asia Pacific, countries do not feel the international threat to collaborate to be respected in the global space sector. A perceived threat though has been established within the region which is causing factions. This regional threat is proving counter-productive in developing regional solidarity for regional collaboration that, in theory, could lead to collective strength on the international level. The global space threat has instead led to collaboration that is highlighting the regional threats and causing regional factions. This has been manifested in the development of two regional space groups: the Asia-Pacific Regional Space Agency Forum and the Asia-Pacific Space Cooperation Organisation (APSCO). APRSAF is driven by the Japanese, Indians and South Koreans while APSCO is driven by China, Pakistan, and Iran. Not only are the driving countries simply different, but they have also been traditionally on opposing political sides on many regional and international issues. For a true regional collaboration, or at least one similar to that in Europe, these serious interregional schisms must be patched before moving forward collectively.

The sense of need in Asia for collaboration is also being overshadowed by the regional factions that have developed. A key advantage for Asia Pacific is that there are several well established space programmes from countries that are showing economic, technical and educational promise. There is much that the countries could exchange at a fair value. For example, Japan has been contributing globally to manned spaceflight for quite some time which could be helpful to the Chinese who could in return contribute knowledge of launchers. Yet this sort of collaboration will not happen for quite some time in the Asia Pacific region because of the regional factions and power struggles.

In Latin America, the threat exists but only for the smaller countries. Brazil and Argentina have space programmes that are developing quickly and are starting to be recognised on the global level. Brazil, accepted widely as the region's most advanced space country, launched a rocket into space in 2004 and Argentina is in the process of building its own launch vehicle for launch by 2013.[143] These countries have already proven that they do not need large-scale collaboration to succeed. The smaller countries, though, see regional collaboration as a way to prevent being left behind in space technology. Given their current status, they have less to lose and more to gain than the larger countries. Furthermore, large-scale collaboration is also not likely in Latin America because the larger countries do not have a need for it. The disparity between the advanced Latin American programmes and the others is such that the larger countries broadly do not see practical value in working with less developed countries. Larger countries would lose programme efficiency without gaining greatly in technological benefits. As mentioned above, this is a perfect example of how regions with space programmes that are not on the same general technical level will not be able in the near future to have large-scale collaboration like that of Europe.

In Africa, the case is similar to Latin America but is more pronounced. The threat exists for all countries even for those that are more advanced in space in the region like Nigeria and South Africa. As a region, there exists an urgency to not be left behind in space and in other fields. When looking at need, again the issue of discrepancy in technological advancement arises. Nigeria and South Africa perceive the threat and see needs to collaborate but are being forced to look outside of the region to countries like China and the United States for support. In 2007, Nigeria launched its first satellite with help from the Chinese, and they hope by 2018 to launch Nigerian-made satellite from Nigeria.[144] Meanwhile South Africa signed a memorandum of understanding in May

1996 with NASA to launch its first South African-made satellite, Sunsat, in 1999 for free.[145] The gaps in space science proficiency between the regionally advanced and less advanced are such that proper motivation for large-scale collaboration does not exist in Africa.

Europe, on the other hand when compared to Asia Pacific, Latin America, and Africa, clearly had both the threat and the need to start its regional collaboration. During the Cold War, the Space Race raged between the Soviets and the United States. At the time, all of the countries on the continent, including traditional global leaders like France and Germany, were faced with the same challenge of rebuilding their infrastructure following World War II. European countries had for many centuries been at the forefront of the world's intellectual and technological endeavors. Now, they stood to fall behind in a key advanced technology – space sciences. To ensure that the region stayed in the same level of competency in space with respect to the Soviets and the Americans, large-scale collaboration made sense. The other enabler of Europe's collaboration in space was that the countries in the region shared relatively similar, advanced technological capabilities and potential for scientific development. The ten founding members of the European Space Agency (ESA) – Belgium, Germany, Denmark, France, United Kingdom, Italy, the Netherlands, Sweden, Switzerland and Spain – all had technological competencies that were worth trading. Just as importantly, there was also a trust that the partner countries would continue to sufficiently invest in long-term space programme enablers like scientific research and education.

Though times have changed, the same threats and needs remain today for Europe. While the Soviet Union no longer exists, Europe still perceives the threat and would like to stay relevant among the world's top space countries like the U.S., China and Russia. Furthermore, the need still exists within European space collaborators. The countries that joined the collaboration have continued to develop their individual space programmes to make them valuable contributions to the European space community. Moreover, the sense of need remains a key driver of the development and structure of the European space community. This is seen, for example, in the way that the ESA requires candidate countries to prove both their financial and technological commitment for a sustained period of time through a three step process before countries are formally admitted. The key third step of the process entails signing a Plan for European Cooperating States (or PECS Charter). This is a five-year programme of basic research and development activities aimed at improving the nation's space industry capacity. At the end of the five-year period, the country can either begin negotiations to become a full Member State or an associated State or sign a new PECS Charter.[146] These continued motivational conditions are what keep the European space collaboration going strong today.

Region	Threats	Needs
Asia Pacific	• Global space threat led to multiple interregional collaboration groups that highlight regional factions and threats • Regional threats are counterproductive to cohesive regional collaboration • Countries do not feel the need to collaborate to be respected in the global space sector	• Several well established space programmes from countries that are showing economic, technical and educational promise • Much that the countries could exchange at a fair value • Need being overshadowed by the regional factions
Latin America	• Some countries, like Brazil, are starting to be recognised on the world scale • Threat exists but only for less space-developed countries	• Space programme disparity is large in the region • Wide disparity dispels need in more advanced countries within region
Africa	• An urgency exists to not be left behind in space and in other fields • Strong threat present for all countries even more advanced in space in the region like Nigeria and South Africa	• Space programme disparity is larger in the region than Latin America • Wide disparity dispels need in more advanced countries within region
Europe	• Large-scale collaboration necessary to ensure that the region stayed relevant in space competency with respect to the Soviets and the Americans at time of founding • Threat from different combination of countries continues today	• Countries shared relatively similar and advanced technological capabilities and prospects at founding ensuring mutual needs • Needs remain today and are built into the growth structure of European space programmes

Fig. 1.9. *Regional threats and needs summary table.*

1.4.4 Cultural conditions for collaboration

These motivational conditions of threat and need are integral for large-scale collaboration. As mentioned, they are relatively objective judgments. Though less tangible and more nebulous, cultural conditions are also very important and can make or break a potential collaboration. Cultural conditions are based in common values and, often directly related to that, historical experiences. Europe's situation is unique because the cultural conditions that contribute so greatly to the space collaboration's functioning are cornerstones of the European identity.

First, Europe has always placed a high value on intellectual endeavors, particularly scientific ones. Since its emergence from the slow Middle Ages thanks to the technological revolution of the Renaissance, Europe has stayed at the top of geopolitics through support from its scientific strength. Further, in Europe the abstract, theoretical, forward-thinking pursuits are appreciated just as much as those that quickly produce practical outputs. Space is arguably the ultimate intellectual endeavor that covers all scientific realms. Astrobiology studies the microscopic origins of life. Satellite development provides mobile phone service on Earth. Astronomy searches for new galaxies so far away from us we will probably never see them. Space provides food for intellectual thought from the micro to the macro scale and from the abstract to the practical, all of which is valued by Europe.

Second, Europe places a very high value on the "community" and working towards helping and protecting the citizens of its community. These values in particular stem from the two World Wars of the 20th century. Partially because of the unique relatively small area of the continent, all European countries felt the devastation of World War I and World War II. A people and infrastructure destroyed, a new European mindset emerged with political structures like the European Union to support it. While the countries in the region are proud of their distinct countries, the idea of working as a community to ensure the security of its people and to propel the region as a whole has become a key part of the modern European identity. This collaborative mindset is important because even while Europe may need to collaborate to be successful in space, they are also driven by a new, culturally-ingrained collaborative nature. Also, space, because of its applications, such as remote sensing for disaster management, farming, drug trafficking and defence purposes, directly supports the security of European citizens. This is to say, there is less difficulty convincing member States to participate in the collaboration, which is not the case in other regions.

The third element of Europe's cultural condition is one that both stems from and supports the first two. Europe merits committing ample and continued funding to collaborative projects where the product may not have an immediate, tangible value and the process may not be the most efficient. While there are other countries that could spend like this, Europe is one of the few regions of countries that can and actually does. An example of this is Europe's Galileo global navigation satellite system (GNSS). The project hopes to provide for Europe an independent GNSS system that will allow European countries to have navigation and communication capabilities. Galileo is currently several years behind and the system's cost has more than doubled from the original estimate in 2003 of 3,4 billion euro to today's 8,6 billion euro estimate.[147] Despite these increases, as of early 2011, Europe reasserted its resolve to complete the system.[148]

It is these elements that make the cultural conditions just right for proper space collaboration in Europe. Not only did these European conditions enable the space collaboration, but also the pursuit of space for Europe contributes to the strengthening of these conditions within Europe. Europe's connection to its identity as a region that places paramount value on intellectual endeavors and sense of community (even perhaps at a high cost) has kept the European space

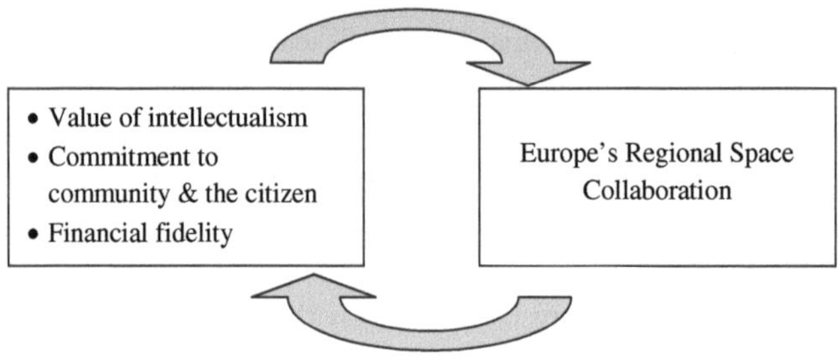

Fig. 1.10. *Cyclical support of European cultural values and its regional space collaboration.*

collaboration going strong. It is a cycle that is self-sustaining which makes Europe's long-term presence in space all but certain.

While these exact cultural conditions are not the only ones that could make regional space collaboration work, they are a particularly strong combination. For the moment, Europe in practice – not just on paper – actually shares these values. It is true that emerging space regions, like Asia Pacific, Latin America and Africa, may have variations of these values or may have individual countries within the region that have these values. For a regional collaboration to work, though, there needs to be a critical mass of countries sharing the values and also practicing them together, not simply discussing them.

1.4.5 Conclusions

Space has always been a scientific realm of which people have been in awe. Now, in addition to fascinating science, the terrestrial applications derived from space are making it a very interesting field in which to enter. Unfortunately, the financial and technological barrier to entry remains high. Collaboration, particularly the European model, may seem like an easy way to tackle these barriers. As demonstrated, though, it takes more than simply a group of countries in a region who all want to benefit from space and who all are willing to combine their space budgets. Countries need to be motivated to work together either out of threat of becoming politically or technologically obsolete and out of technological need. Additionally, there should be shared regional cultural conditions that support not just collaboration but space collaboration and its unique intellectual, strategic and financial challenges. The cultural conditions are not just a "nice to have." Arguably it is these cultural conditions, so ingrained in Europe's identity that has continued to propel Europe's space collaboration.

At this point, these conditions simply do not exist for the emerging space regions of Asia Pacific, Latin America and Africa. While collaboration seems like a nice idea politically, at the moment, it is not worth it for countries in these regions to invest the financial or political resources necessary for regional large-scale collaboration. Indeed, as described in the NRC report, collaborative programmes should not be undertaken because of "imposed merger of technical requirements for political reasons, external hopes for cost savings, the addition of unfunded mandates, or directed interagency collaboration for the sake of collaboration."[149]

Because cultural conditions are harder to change, these regions should work on the motivational conditions. As explained, the asymmetric space programmes within the respective regions are a key roadblock. This, perhaps, is where the energy must be invested particularly in Latin America and Africa. Emerging space countries should look to obtain a base level of space technical capacity before the regional collaboration is reexamined. Asia Pacific is a more complicated situation as the interregional threat may simply remain a roadblock for a long time.

What goes without question is that the global development of space will move forward. As more players emerge the question of regional and even global collaboration will only become more complicated. What will remain, though, are the basic motivational and cultural conditions needed for collaboration in space that have made Europe the best model to date.

[133] Space Foundation. *Globalization of Space Draws New Market Entrants.* 23 August 2010. 7 September 2010. *http://www.spacefoundation.org/news/story.php?id=1003.*

[134] European Space Agency. *What is ESA?.* 6 August 2010. 22 January 2011. *http//www.esa.int/SPECIALS/About_ESA/SEMW16ARR1F_0.html.*

[135] Committee on Assessment of Impediments to Interagency Cooperation on Space and Earth Science Missions; National Research Council. *Assessment of Impediments to Interagency Collaboration on Space and Earth Science Missions.* Washington, D.C.: The National Academies Press 2010: 1–4. 12 February 2011. *http://www.nap.edu/catalog/13042.html.*

[136] *Ibid.*: 31.

[137] *Ibid.*: 31.

[138] *Ibid.*: 10.

[139] *Ibid.*: 2.

[140] Committee on International Space Programs, National Research Council, and European Space Science Committee, European Science Foundation. *U.S.-European Collaboration in Space Science.* Washington, D.C.: The National Academies Press, 1998. 12 February 2011. *http://www.nap.edu/catalog.php?record_id=5981.*

[141] "ISRO Rockets into Higher Orbit with 35% Hike." *Times of India.* 27 February 2010. 12 February 2011. *http://timesofindia.indiatimes.com/home/union-budget-2010/Isro-rockets-into-higher-orbit-with-a-35-hike/articleshow/5622518.cms.*

[142] Lewis, Leo."South Korea joins developed nations' club but may pay a price." *The Sunday Times.* 23 September 2008. 12 February 2011. *http://business.timesonline.co.uk/tol/business/markets/article4810718.ece.*

[143]"Argentina developing satellite launcher; ready by 2013 in Puerto Belgarno." *Merco Press*. 16 August 2010. 28 March 2011. *http://en.mercopress.com/2010/08/16/argentina-developing-satellite-launcher-ready-by-2013-in-puerto-belgarno.*
[144]Isoun, Turner, "Space Landings." *Nature*. 30 October 2008. 12 February 2011. *http://www.nature.com/nature/journal/v456/n1s/full/twas08.35a.html.*
[145]University of Stellenbosch. "Sunsat FAQ" 7 Jun. 1999. 12 February 2011. *http://research.ee.sun.ac.za/sunsat/.*
[146]European Space Agency. "Plan for European Cooperating States." 18 January 2010. 12 February 2011. *http://www.esa.int/SPECIALS/PECS/SEMXSFZRA0G_0.html.*
[147]Cendrowicz, Leo. "Galileo: Europe's Satellite Scheme is Lost in Space." *Time*. 25 January 2011. 28 March 2011. *http://www.time.com/time/business/article/0,8599,2044300,00.html.*
[148]"The Mid-term Review of the European Satellite Radio Navigation Programmes Galileo and EGNOS: Questions and Answers." 19 January 2011. 12 February 2011. *http://www.iewy.com/14532-the-mid-term-review-of-the-european-satellite-radio-navigation-programmes-galileo-and-egnos-questions-and-answers.html.*
[149]Committee on Assessment of Impediments to Interagency Cooperation on Space and Earth Science Missions; National Research Council. *Assessment of Impediments to Interagency Collaboration on Space and Earth Science Missions*. Washington, D.C.: The National Academies Press 2010: 34. 12 February 2011. *http://www.nap.edu/catalog/13042.html.*

Technology

1.5 European identity, technology and society. The case for space

Stephan Lingner

1.5.1 Point of departure

Science and technology are clearly areas that have contributed to the notion of Europe as they are integral parts of its cultural heritage. The question is whether European culture was *determined* by technology development or not. The immediate answer is "no", because we generally cannot state a strong causal relation between technological and social development by simply assuming a social shaping role of technology. On the other hand, this relation cannot be reduced to a mere co-incidence of technology and society without any mutual constraints. The truth lies in between in terms of a weak coherence of both with respect to a conditional relation between technology and society. Technology is clearly – but not exclusively – a cultural practice and as such it might push societal development among other non-technological practices. On the other hand, technology is governed by society in nearly every phase of its development and diffusion, e.g., by funding research and specific projects or by regulating certain applications. The resulting *co-evolution of society and technology* means that European culture and identity cannot be simply shaped by ambitious technology development – but may gain from corresponding incentives, when development conditions for technology and society are integrated appropriately.

This holds especially true for spaceflight, which is an exceptional large-scale endeavour with corresponding strong implications for society, culture and identity. On one hand its visibility and fascination inspires the public, politics and researchers more than most other fields of human action. On the other hand large-scale endeavours like human spaceflight entail large expenses by governmental bodies. Therefore, corresponding decisions have to be legitimated with respect to limited public funds and competing demands from other actors, as well as with respect to the aims, benefits and meaning of spaceflight. Related debates and *acceptance problems of spaceflight* highlight its specific ambivalence in modern societies, which requires a rational assessment of related options, potentials and risks.

Against this background, on 15 and 16 October 2009 the European Commission and the European Space Agency (ESA) together with representatives of the relevant industry held a joint conference to unravel the specific "*Ambitions of Europe in Space*". Among other findings, one summarising thesis of the organisers claimed that: "Finally, space exploration (is) to be integrated in the broader prospect of economic development". This utilitarian perspective is generally plausible and legitimate but may prove to be too narrow as regards the comprehensive cultural aims of spaceflight that mainly unfold beyond the utilitarian horizon.[150] Therefore, the ambitions of Europe in space might be discussed in a broader context especially concerning their impact on European identity. Corresponding argumentations will be addressed in the following sections.

1.5.2 The notion of Europe

Throughout the past several decades, the European idea has grown increasingly towards a common *European identity*. This process has been accompanied by the establishment of a joint economic and research area, the disappearance of border controls between the Member States and the introduction of a common European currency. Together with European regulatory competences, these fields now cover the main areas of relevance for public life in Europe. Further consolidation of the European citizenship may be expected through the results of the Bologna process in academia and by the implementation of the Lisbon strategy, which aims – among others – at greater sustainability, security and coherent innovation paths within Europe, its regions and sectors as well as at the advancement of capacities to compete with strong global players, like the United States of America. Nevertheless, the eastern enlargement of the EU towards the western border of Russia and the issue of whether Turkey should belong to the European Union or not, raise the question as to on which common characteristics or criteria a corresponding self-conception of "Europe" might be reasonably based. A more detailed analysis will be given in the next few chapters.

1.5.2.1 Europe as a geographical figure

The geographical description of the European continent dates back to the 6th century BC as part of the ancient world, which was restricted to the known landmasses surrounding the Mediterranean Sea. Figure 1.11 gives an impression

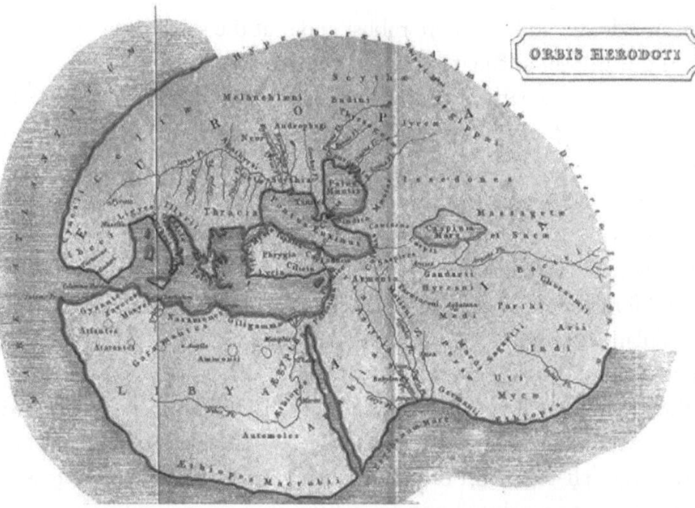

Fig. 1.11. *Europe and the ancient world at Herodot's times (source: Butler 2005).*

of its outline according to the state of antique cartography.[151] In those days, the shape of the known world was described by a quasi-circular speculation with the centre in the eastern Mediterranean. From there, the continents were simply defined by dividing the "world's disk" into three sectors. The northern sector was then labelled "Europa" according to the historical name of a northern Greek province in Herodot's times.[152] This rather schematic figure stayed astonishingly stable up to modern times – still omitting the heartland of today's Turkey. Nevertheless, it seems questionable whether this *arbitrary geographical perspective* could reflect a common European identity.

1.5.2.2 Europe as a political entity

The ancient geographical notion of Europe evolved as political realities changed throughout the history of the continent. Its ancient cultural centre shifted from Greece to Rome and moved further to the north in the middle ages, without changing its shape significantly however. Consequently, the Congress of Vienna in 1815 approved geographical Europe as a whole – irrespective of its internal reorganisation. This conventionalist but still arbitrary approach had its practical value as long as obstacles from inconsistencies could be kept apart by low expectations towards Europe as a normative figure and by the reality of superior national powers. However, the ongoing European process and the global extension of European "ideas" call for a more robust and – at the same time –more flexible orientation towards an adequate understanding of "Europe".

The question as to whether today's European institutions already reflect these requirements or not, can be answered differently as shown by three relevant examples:

The launch of ESA goes back to the "Space Race" after the Soviets got first access to space with their Sputnik probe. The ESA programmes therefore aimed at an independent technological competence in Europe and attracted a growing number of Member States as well as Canada as an active Associated Member. However, Poland is only a Cooperating State since 2008, while it ought to become a full Member State with respect to its cultural roots.

Even the expanded European Union (EU) still lacks members such as Switzerland, which refrains from joining the EU for reasons of economic and political autonomy. This self-exclusion of a European country may be seen as arbitrary again with respect to a cultural notion of Europe.

Only the Council of Europe seems to address the appropriate values and countries with regard to the identity problem by emphasising the *common cultural heritage* of the European countries. Consequently, states such as Russia and Turkey also belong to this community together with other more familiar European Member States. Principally, this institution might therefore serve as a model for the adequate representation of Europe, although the political practices of several eastern Member States may conflict with the shared human and democratic values of the community.

1.5.2.3 Europe as a concept

As demonstrated above, the classical notion of Europe suffers from geographical arbitrariness, related conventionalism as well as from a one-sided orientation as a consequence of technological, economic or political motivations. Nevertheless, culture as a whole counts in the end when considering European identity, which includes far more than technological, economic and political practices. Recall that Europe's tradition goes back to Antiquity, when rational thinking, science, culture and trade were established. Figure 1.12 shows an ancient portrait of Aristotle, who was significantly involved in early conceptions of rational reasoning, science and politics, which heavily influenced further societal developments on the continent and which are mostly still valid today. Related recent challenges continue to necessitate rational reflection on the technological options, risks and choices faced by modern European societies and on democratic procedures, legitimate decisions and enterprises built thereon.

Fig. 1.12. *Bust of Aristotle, Greek philosopher (4th century BC) (source: Wikimedia Commons).*

In sum, Europe is historically rooted on the continent, but not closely restricted to it. Its openness is solely framed by the concept of rationality. Europe had started with an idea of *ratio*, which might not have turned out to become an ideal. In this sense, rationality might be seen as a *unifying paradigm* for the European identity.

1.5.3 The role of rationality

Rationality is related to a significant number of values that may be attributed to European identity: (1) Rationality is a key to the truth, while aiming at the proof of consistency of assumptions and theses. The "programme of rationality" results in the enlightenment and *refinement of world views*, which is not trivial as it is the prerequisite for human autonomy and emancipation from incompetence according to the philosopher Immanuel Kant.[153] (2) In this sense, rationality enables progress through scientific verification and validation of models and systems and therefore, it contributes to the *culture*

Fig. 1.13. *Exploring new worlds (Viking spacecraft over planet Mars, 1976) (source: NASA).*

of innovation. (3) Rationality is also a means for efficient allocation of limited resources or time in terms of cost-benefit analyses of financial investments and through comparative calculation of external costs as trade-offs for the environment and the society. In this way *it contributes to welfare.* (4) Moreover, a consensus-building potential of rational argument can be stated with respect to expectations and the need for reasoning certain claims or measures in public policy. Rationality is thus an adequate means for acceptable and *peaceful conflict management* in modern societies. (5) Summing up, rationality might be seen as a sound common or even *universal cultural practice* serving as an identity-building element in Europe and beyond. Residual problems of rational acting, caused by ambiguities in ultimate goals, uncertain side effects and divergent expertise leading to experts' dilemmas should not be denied as objects of so-called "post-normal science".[154] However, defining tolerable loads and acceptable standards is again a matter of rational (re-) evaluation in-progress.[155]

1.5.4 The case for rational technology assessment

As shown above, modern technologies – such as nuclear power systems – and their outcomes are often ambivalent for society and the environment. Technology assessment will therefore have to (a) weigh the respective chances and risks of any option, (b) compare different risks from different options to be selected, and (c) evaluate actor-specific subjective preferences against possibly violated universal ethical values. Idealised rational decision processes will thus assess technology developments at both the means *and* aims levels. The latter is directed towards the normative appropriateness of major goals of certain technical developments and the prioritisation of possibly conflicting internal and external aims. The former addresses the coherence or consistency of related measures, strategies and alternatives in efficiently meeting goals, through rational analysis of the necessity and sufficiency of supposed means. Technology assessment is therefore a means of finding rational choices and not the phantom of "technology arrestment" which some sceptics fear(ed).[156] It supports the rational governance of complex, uncertain or ambiguous issues, raised by modern technology, that challenge today's societies. Technology assessment is thus a consultancy service to the responsible actors for science-based policy advice to find reasonable and legitimate solutions.

In this sense, the necessity of rational technology assessment becomes exceptionally evident in the challenging case of spaceflight and related technology developments. It resembles a special case with respect to corresponding large-scale infrastructures to be built-up, their impressive visibility and their huge expenses from public funds, which have to be legitimated. Technology assessment becomes here a task of framing decisions on the development of costly infrastructures while explicitly considering related chances and risks as well as the specific meaning of spaceflight to humankind. Corresponding evaluations have to be conducted at different hierarchical levels, starting with cost-benefit analyses, moving to assessment of the adequacy of means and strategies to meet the goals of space activities and then towards the proof of legitimacy of certain exploration goals with respect to possibly endangered universal norms and to an assumed ultimate objective of spaceflight. The latter should be plausible but it principally cannot be reasoned in itself.[157] In this sense, *culture* – as it embodies the essence and condition of humanity – might be seen as an ultimate goal of space exploration.

1.5.5 Spaceflight – a cultural option

Spaceflight might be therefore seen as a cultural effort and future option of humankind. Its manifold aspects can be divided into separate fields, which are

Table 1.3: *The cultural dimension of spaceflight (own source)*

Cultural Aims	Options and Perspectives	Yields (short-/long-term)
Enhancement and expansion of the human cultural sphere; liberation from limiting borders	Space exploration and human presence in space	
Polycentric world and multi-lateral political culture on Earth	Peaceful international cooperation; plurality of spacefaring nations and balance of power; broad participation in new options in/from space	
Cognition and development	Scientific knowledge; extended world views; education; new technologies	
Utilisation of space assets	Returns for environment and security; added economic value; entertainment	

outlined in Table 1.3. Most of the relevant categories plot beyond the short-term perspective of concrete paybacks. Instead, they foster human cognition – which is clearly a cultural aim – and human development. The latter will even reach beyond levels of technological development, because corresponding efforts (complex missions; space debris mitigation) are also needed for international cooperation, thus probably enhancing societal developments towards a distribution of power and peaceful participation of space-faring nations on Earth. In the long-term, spaceflight might also contribute to the liberation of humankind from terrestrial borders, enabling the extension of human activity to extraterrestrial space as well as to the establishment of a future cosmic existence of human life.[158]

1.5.6 Conclusions and prospects

Spaceflight is often discussed in too narrow utilitarian or even economic categories. Its potential for peaceful international cooperation, both on Earth and in space, should be borne in mind. Overall, spaceflight is clearly *a matter of culture*, thus expanding any perspective restricted to short-term benefits. The objective of cultural development gives reason – but not necessarily priority – to spaceflight as a *rational human endeavour* even beyond the limited horizon of cost-benefit analysis. This view also takes into account the notion of rationality as a paradigm and unifying heritage of Europe, which should be seen as a concept and idea rather than a geographic entity. In this sense, spaceflight should be more tightly aligned to the European cultural tradition of rationality – ranging far beyond the horizon of economic scales while strengthening a common identity.

The prospects for corresponding endeavours are currently not too bad: the "Augustine Report" to the U.S. White House on 22 October 2009 recommends that NASA should strongly cooperate with other space-faring countries in ambitious missions beyond low Earth orbit, partly due to budgetary constraints.[159] This proposal was strongly supported by the French Research Ministry, which announced a complementary European initiative towards international space exploration of the solar system.[160] With this perspective, near-Earth space could increasingly be explored by private investors, who are already actually planning related space travel (Virgin Galactic and SpaceX programmes). A Nature survey showed that the spirit of, and inspiration from, space exploration should not be underestimated, which might generate public support for corresponding initiatives.[161] In this sense, the exceptional visibility especially of spaceflight can be seen as a resource that should be better harvested in the future – also to the benefit of the cultural development of humankind, at least from a "European" perspective.

[150]Moreover, the "utilitarian fallacy" in reasoning about space activities might also raise false expectations and respective disappointments, which – in the end – might even affect the spirit of spaceflight: see Lingner, Stephan. "Human Spaceflight as a Matter of Culture and National Vision." Humans in Outer Space – Interdisciplinary Odysseys. Luca Codignola and Kai-Uwe Schrogl (eds.) Wien New York: Springer 2008: 175–181.

[151]See Butler, Samuel. "The Atlas of Ancient and Classical Geography." 21 November 2005. The Project Gutenberg EBook. EBook #17124. 20 October 2010. ***http://www.gutenberg.org/files/17124/17124-h/17124-h.htm.***

[152]See Cobet, Justus. „Wie weit reicht Europa und was heißt Europa? February 1998. Newsletter no. 8. Europäische Akademie Bad Neuenahr-Ahrweiler GmbH. 20 October 2010. ***http://www.ea-aw.de/fileadmin/downloads/Newsletter/NL_08_021998.pdf.***

[153]See Kant, Immanuel. „Beantwortung der Frage: Was ist Aufklärung?" Berlinische Monatsschrift (Dec. 1784): 481–494. 20 October 2010. ***http://www.uni-potsdam.de/u/philosophie/texte/kant/aufklaer.htm.***

[154]See Funtowicz, Silvio, and Jerry Ravetz. Post-Normal Science. Science and Governance under Conditions of Complexity. Interdisciplinarity in Technology Assessment. Michael Decker (ed.) Berlin Heidelberg New York: Springer 2001: 15–24.

[155]See Carrier, Martin. "Business as Usual: On the Prospects of Normality in Scientific Research." Interdisciplinarity in Technology Assessment. Michael Decker (ed.) Berlin Heidelberg New York: Springer 2001: 25–32.
[156]See "The Debate over Assessing Technology: Congress wants to set up an office to evaluate the impact of New Developments." 8 April 1972. Business Week. 20 October 2010. *http://www.princeton.edu/~ota/ns20/ota72_f.html.*
[157]Gethmann, Carl Friedrich: „Letztbegründung." Historisches Wörterbuch der Philosophie. Joachim Ritter, Karlfried Gründer and Gottfried Gabriel (ed.). Vol 5. Basel: Schwabe 1980.
[158]Mendell, Wendell W. (ed.): "Lunar Bases and Space Activities of the 21st Century." Houston: Lunar and Planetary Institute, 1985. See also: Stöffler, Dieter: "Die Erde in eine offene Welt verwandeln." Wirtschaft & Wissenschaft Mai 2/99 (1999): 38–47.
[159]„Flying high." The Economist, 12 September 2009: 83–84.
[160]"L'Europe regarde vers la Lune et Mars." Le Figaro, 24 October 2009.
[161]Monastersky, Richard. „Shooting for the Moon." Nature 460 (2009): 314–315.

Industry

1.6 Does the European integrated space industry contribute to a new European identity?

Mathias Spude

„Car nous autres Européens, nous devons êtres considérés comme des voyageurs embarqués sur un seul et même navire“[162] (Comenius, Prague 1645). More than 350 year ago, Comenius the patron of the European school exchange programme, put into words what European integration and identity is all about: We are deemed to sit in the same vessel and must make the best of it.

1.6.1 Introduction

Defining European identity is a rather challenging mission these days, be it politically, or from a cultural or an industrial perspective. There is no other topic in European matters that is as vague, unclear, and ambiguous as the issue of European identity.[163] The evaluation of identity is difficult and full of misunderstandings, because identity can be understood as fact, as appearance, and as emotion.[164] It is to the credit of the European Institute for Space Policy (ESPI) to have put this challenging topic on the agenda in the context of European space activities.

Astrium, the EADS space branch, was created in 2000 and is the emblematic European integrated space industry company as it operates in five European countries: France, Germany, the United Kingdom, Spain and the Netherlands. Astrium is the only space company covering all space systems, ranging from launch vehicles to satellites of all kinds, space systems for human spaceflight and space-based services. Hence, it is desirable and useful to investigate if Astrium as the European integrated space company is helping to engender a sense of European identity.

The approach to shed light on the concept of European identity by linking economy and identity may seem daring. Nevertheless, European integration started in the economic field and today Europe could not be conceived of without its great economic power. Therefore, an integrated European company, such as Astrium in the space sector, seems to be a very valid object to investigate the generation of European identity by commercial and industrial companies.

There is not a lot of research and literature on the ability of industrial structures to generate European identity. Without such reference points, this investigation will take advantage of such academic freedom and will follow two paths: on the one hand "corporate identity" will be defined and an analysis will be made of whether and how the Astrium corporate identity helps to create a European identity. On the other hand, existing definitions of European identity from other domains will be examined, and the question of whether an integrated industrial concern such as Astrium has the potential to create a European identity will be validated by way of analogy. The conclusion will summarise the results.

1.6.2 Space activities in Europe

1.6.2.1 The institutional framework

Why should we look at the institutional space framework in Europe when trying to identify a European identity generated by the space industry? Most space based infrastructures and certain technology developments are financed by public institutions, reflecting the fact that space activities enable the implementation of governmental and public tasks such as fundamental research, scientific activities, mobility, security, international cooperation and many not for profit activities. Thus public funding strongly dominates the turnover of the European space industry including Astrium, whereas the commercial part of the turnover is lower (between 50–60% of the institutional turnover)[165] and subject to market variations.

The European Space Agency ESA is Europe's cooperative organisation for space research, technology and space applications.[166] The ESA Convention states that "the purpose of the Agency shall be to provide for and promote . . . cooperation among the European States . . . ".[167] ESA elaborates and implements a long term European space policy, European space activities and programmes, and implements an industrial policy.[168] ESA programmes cover European independent access to space with the Ariane 5 launcher, European participation in the

International Space Station (ISS), the observation and exploration of the Earth, the terrestrial environment, the exploration of the solar system and universe. ESA programmes cover also everyday applications, such as navigation and telecommunication programmes including all necessary technologies, as well as the European spaceport, the ESA laboratories and test facilities and the European astronaut corps.

Through ESA development activities, several European space organisations in the operations and services fields were created, such as Arianespace, EUMETSAT, and EUTELSAT, which also contribute to creating a European identity through space activities; be it through their character as international organisations and the political will behind their creation and existence, or through their European activities and the appearance of their products and services.

The European Union (EU) has emerged as a new space stakeholder in Europe during the last decade. The EU role in space activities is threefold:

1. The EU has own responsibility for space policy since the Lisbon Treaty came to force;
2. The EU is financing space infrastructure programmes: the new European satellite navigation system Galileo is mainly financed by the EU with ESA managing the implementation;
3. The EU is financing space application programmes: the Global Monitoring for Environment and Security (GMES) programme with the purpose of facilitating the tailor made application of space based data is financed by the EU (the satellites being financed by ESA and national states).

1.6.2.2 A European identity and national identities via institutional space activities?[169]

Looking at this institutional framework it is fair to state that ESA intended and continues to contribute to the idea of a European identity. This is proven by the sheer success, great public attention and immense political support linked to almost 40 years of ESA programmes and projects.[170] The EU GMES and Galileo programmes are driven and perceived as joint programmes in the interest and for the benefit of Europe, thus also creating European identity.[171] This way of looking at the institutional space activities in Europe can also be based on the trans-utilitarian aspects. These are not economically measurable, but they describe substantial benefits: They cover aspects like satisfying the desire for discovery and curiosity, enlarging culture and the "Weltbild", creating emotions, generating inspiration, generating prestige and leadership and influence.[172] These aspects also engender elements of European and national identities.

At the same time there are strong and well developed national space identities in Europe. In particular, for more than 40 years France, Germany, Italy, the United Kingdom and Spain have been running their own national space programmes and activities in parallel to their ESA participation.[173] Being an ESA Member State does not exclude the national dimension: one of the fundamental principles of ESA is the geographical return, safeguarding that the financial input to ESA flows back to the state by way of contracts to its industry and research organisations.[174] It is fair to state that this geo-return rule demonstrates the duality of European and national space identities. And one could even go down to the next level, i.e. the regional level. In France and Germany in particular, the national space research centres, universities and space industry are geographically spread within the country. Taking into account the political support at regional level and the public support, this can also be seen as a regional space identity in Europe.

Looking at the institutional level, one could say that there are European, national and regional identities existing in parallel.

1.6.3 A "de facto" European identity created by Astrium

Even lawyers take the normative power of the factual for granted.[175] Hence it makes sense to start by looking for a European identity as a result of the sheer existence of Astrium. Astrium is the EADS space division.[176] EADS including its space division Astrium evolved as a result of the political and industrial will to create a worldwide competitive European industrial entity.[177] This can be interpreted as a first indication of a European identity.

The people working for Astrium may be considered to be an active factor in generating European identity. Today, Astrium has a workforce of almost 16,000 employees in France (44%), Germany (27%), the UK (23%), Spain (5%) and the Netherlands (1%). The fact that these people have been working together successfully in one company every day for almost a decade on the same space programmes and projects at least indicates but probably even reflects a growing European identity.

The third indication of a "de facto" European identity is the space activities where Astrium has extensive experience as an industrial prime contractor: satellite systems, payloads and equipment for civil and military applications, launch capabilities, orbital systems and manned space activities, and a wide portfolio of space-based services.[178] It can be assumed that a company that develops, integrates, launches and operates all these European space projects has a significant role in generating a European identity.

The fourth indicator of a European "de facto" identity is the economic power of Astrium: Astrium's turnover has increased significantly over the years: from 2,535 billion euros in 2000 to 4,289 billion euros in 2008, with an order backlog of 4,826 billion euros in 2000 to an order backlog of 11,035 billion in 2008. In economic terms Astrium is the No. 1 space company in Europe. As the main industrial space stakeholder, Astrium clearly plays its part in creating a European identity.

1.6.4 European identity created by Astrium corporate identity and culture

Based on the analogy of every human being having his/her own personal identity, "corporate identity" is the notion of an enterprise's identity and personality, manifesting its uniqueness. The objective of corporate identity is the sustainable evolution of the company.[179] The classical definition of corporate identity is: the strategically planned and operatively implemented presentation and behaviour of a company on the basis of a company philosophy, long term company objectives and a well defined "want to be" image – with the aim to present all activities of the company in a coherent way to the inside and outside.[180] Corporate identity is the coordinated use of corporate behaviour, corporate communication and corporate design (visual presentation) inside and outside of a company. Important additional factors of corporate identity are the company's history, its organisational structure, the company vision and its guiding principles. What corporate identity does Astrium have and does it constitute a European identity?

1.6.4.1 Astrium guiding principles

According to the Astrium guiding principles, Astrium is the only European company mastering satellites, launchers and services. It is the Astrium mission to "develop and successfully serve customers and markets . . . to create value for customers, run operations on a multi-country basis, drive European integration . . . and cooperate to Astrium transversal projects".[181] These principles make it quite clear that the active generation of a European identity is one of the core aspects of Astrium's guiding principles. This is reinforced by the use of notions such as "European integration", "multi-country basis" and "transversal projects".

Astrium at the same time sets itself the objective of reinforcing Astrium's position as a national player in each home country.[182] It is the guiding principle of Astrium to implement the potential of an integrated industrial structure at European level with a European identity, and at the same time to maintain a national identity. This reflects fully the scope of institutional space activities and the European and national identity they simultaneously create.[183]

1.6.4.2 Astrium values

What are Astrium's company values and do they reflect the creation of a European identity? The value "leadership" stipulates that Astrium's ambition is to be the European leader in satellites, launchers and services business. The value "team spirit" stipulates that Astrium acts as one team and promotes exemplary "working together" practices throughout the company. At the same time, the Astrium value "people" states that Astrium values excellence and makes the most of its cultural diversity.[184] When looking at these values, the facilitation of and contribution to generating a European identity are evident and core to Astrium.

1.6.4.3 Astrium structure

The Astrium structure reflects the fact that European identity is a guiding idea. Astrium is organised along three dimensions: business units, high level national representatives and corporate functions. The Astrium business units serve customers and markets while the high level national representatives establish Astrium as preferred national player in each home country, relying on its multi-national base. And the corporate functions build a coherent and integrated Astrium approach through optimised integration.[185]

Astrium is organised operationally in three business units: Astrium Space Transportation, Astrium Satellites and Astrium Services. These business units have a profit and loss responsibility and are operationally self standing. They are organised across borders, and the daily business, management, leadership and responsibility are fully transversal.[186] This reflects the idea of a European identity. The same goes for the central functions of Astrium (strategy, financial controlling, human resources, institutional relations, and communications & public relations) which are fully centralised.

The Astrium paradigm of working together is well implemented through the Astrium European Work Council, where employee representatives from all Astrium countries are involved through participative elements in decision making.

At the same time, there exist five national legal Astrium entities comprising all national employees and assets in each of the five home countries: Astrium SAS in France, Astrium GmbH in Germany, Astrium Ltd in UK, Astrium SL in Spain and Dutch Space BV in the Netherlands. In each country, a "High Level National Representative" represents all national employees, assets and activities towards the government, the workers' representatives and towards media and the public.

It cannot be denied that such an organisation is complex and challenging on a day to day basis, but this is the only way to strive for European integration. Here we find the analogy to the creation of identity by the institutional space stakeholders in Europe.[187] Just as there is a European identity via ESA, the EU and other European institutions in parallel to the strong national institutional identities of the European space faring nations, we find the same with Astrium. Astrium on the one hand strives to implement the idea of European identity and at the same time allows for a national identity.

1.6.4.4 Astrium history

Astrium, being a fully integrated European space company, is the result of a considerable national and European process of industrial concentration. In the early 1990s, the former separate and competing companies of Germany Dornier, MBB and ERNO were merged into the national German Dasa. Similar developments happened in France. As the result of political will and decisions, Matra Marconi Space and Dasa Space Infrastructure were merged in 2000 into "Astrium 1", and in 2003 merged with Dornier Satellites and Aerospatiale into today's Astrium. In 2005 Dutch Space was integrated.

Looking at this quite impressive process of industrial concentration, it becomes obvious that the idea of European identity was a driving force. To create one single integrated company out of several predecessors is an exercise that is also driven by the motivation to create a new – European – identity. Even though this history disappears with the change of generation of Astrium employees, it is a factor that influences the speed of generation of a European Astrium identity.[188] The other main driving force to create a European integrated company was to create an industrial counterpart to the strong U.S. space companies dominating the market. This motive can be seen as pro European too, as the underlying motivation was to bring economic profit to Europe as a whole.

1.6.4.5 Astrium corporate behaviour

Corporate behaviour is the way the employees of a company behave towards customers, partners, and suppliers, towards the government and public, as well as among themselves. Corporate behaviour is the most effective instrument in implementing corporate identity, as it reflects the company's credibility.[189] The Astrium values mentioned above are put into concrete terms by a so called set of behaviour that contains principles of behaviour that are part of the yearly objective setting with the employees. It would be beyond the limit of this investigation to fully analyse the corporate behaviour of 16,000 employees. Nevertheless a few examples and aspects should be mentioned. Astrium regularly carries out customer surveys, in particular among the institutional customers ESA, DLR and CNES. The results of these surveys show that Astrium representatives are respected for their space know-how and expertise, and that they are part of one company. Of course, there are critical remarks by the customers, such as complex decision making in Astrium, but these remarks reflect that Astrium is regarded as one company. The centralised Astrium corporate functions have the mission to also align the messages and behaviour of Astrium towards governments and institutions, reflecting again the aim for one company identity. Another factor in creating a common corporate identity is the joint representation of projects by Astrium employees (from several nations) towards the customer. As another example, Astrium regularly organises supplier conferences in order to maintain communication to suppliers and partners in a sustainable way. These conferences are very well received by Astrium suppliers and demonstrate the Astrium corporate identity.

Concerning the behaviour of Astrium employees among themselves, a survey is ongoing at the time of writing this paper. Hence, the results cannot be given here. Notwithstanding, it can be stated that creating a common Astrium identity among 16,000 employees remains a challenge. On the one hand, the centralised Astrium corporate function "human resources" has the mission and facilitates the improvement of working together and implements training and interaction accordingly. In this context the Astrium managers are evaluated yearly by their hierarchy with regard to their behaviour inside and outside of Astrium. This is important and part of the manager's yearly salary is linked to this evaluation. Additionally the company structure encourages and to a large extent demands that employees work together across borders – in the widest sense – and all this for almost a decade. On the other hand, people have their own history and professional experience and it is always a challenge to get people with separate cultures to work together. The result of an interesting investigation is that it is a continuous process for the company, its management and the managers to understand, accept and manage separate cultures in the best interests of Astrium.[190]

1.6.4.6 Astrium corporate communications

From the very beginning, Astrium communications were centralised as a corporate function. Consequently, all preceding national and divisional communication organisations were merged into one single corporate Astrium function. Taking into account the Astrium history with several competing companies that had been cooperating on selected projects but never as a single company, it was a strategic choice to centralise corporate communications as an essential tool for creating and conveying an Astrium corporate identity. The Astrium communication directorate's mission is to define Astrium's global communication strategy, to manage the information and public relations policy and to implement the communication plan and operations.[191] The Astrium communication directorate's organisation reflects the target groups: the internal Astrium population (internal communications function) as well as the external side (media relations, public relations, exhibitions, and editorial functions). On the basis of a yearly communication plan, all communication and public relations activities are implemented. All actions reflect the objective of communicating the Astrium image and identity as a European integrated company.[192] Nevertheless, the success factor and limit of corporate communication is credibility. Hence, in parallel to success stories, critical aspects and points for improvement are communicated too.

1.6.4.7 Astrium corporate design

In order to be comprehensive, it should be stated that Astrium has a complete set of corporate design/visual identity rules, that need to be and are respected whenever Astrium presents itself to the inside and outside. The power of visual identity is obvious to everybody when thinking of the powerful brands we see everyday.[193]

1.6.5 Creation of European identity in other terms?

Examining Astrium's potential to generate a European identity only by analysing the corporate identity dimension would not meet the complexity of this issue. It is important to also look across the boundary of the corporate world and see how other domains of public life define European identity, and to apply these ideas to Astrium.

1.6.5.1 European identity in political terms

European identity in terms of politics is based on two premises: (1) the consciousness of its inhabitants to belong to a common regime/body politic, and (2) that the citizens accept the political project of their common regime/body politic.[194] What would this mean for the integrated Astrium space company? After almost 10 years in existence, all Astrium employees are aware and conscious that they belong to Astrium as an integrated company. Each employee knows that he/she works for this company. Thus, the first criterion is fulfilled. The second part of "accepting the common regime" i.e. Astrium in this case, is far more difficult to acknowledge or deny. The fact that Astrium employees accept belonging to an integrated Astrium can at the end of the day only be evaluated by means of an employee survey. Such a survey is currently ongoing, but results are not yet available. Nevertheless, all informal feedback activities have indicated that it takes time to get a workforce of 16,000 to buy into the concept of an integrated European company. Thus, it can be stated that Astrium is creating a European identity based on the consciousness of its employees, with the soft factor of accepting that this identity is a process and is evolving over time.

1.6.5.2 European identity in the light of national identities

During the creation of nation states in the 19th century, three core factors were identified that were essential in the social process of creating a political identity: (1) a common system of education, (2) the military draft, and (3) a well established national public domain.[195] Of course it is somewhat daring to use these three factors in the context of Astrium and its potential to create a European identity, but who does not dare, does not win . . . Concerning 1) Astrium employees do not go through a common system of education as they are recruited from more than 10 nationalities. But all newly recruited Astrium employees receive similar company information, there is a system of "godfathers and -mothers" who coach new employees, and the managers have to go through similar training courses. Hence, the first steps of a common Astrium education are evolving. Concerning factor 2 (military draft) it is of course obvious that this does not exist in Astrium! But every Astrium employee – throughout all hierarchy levels – has an annual interview with his boss to review the performance of the last year and to set the objectives for the coming year. When being promoted, it is mandatory to have participated in special training courses. Factor 3 – a well established public domain – is easy to confirm with regard to Astrium. The high number of articles in the media and discussions in the space community, as well as the constant discussions about identity and future of Astrium among the employees together with the activities in the Astrium internal communication, prove fully that there is a well-established public domain. Hence, from this point of view, the creation of a European identity by Astrium is evolving.

1.6.5.3 European identity by participation

In the political context, European identity may be promoted by participation and participative governance. On the one hand it is not conceivable to manage a 16,000 employee company by means of participation; on the other hand, Astrium employees are highly skilled and trained, with some 70% having college and high school diplomas. There are also participative initiatives such as "Top 80 Manager" meetings, general management conferences etc. where senior management meets the line managers to discuss matters of strategy, objectives, and other relevant issues. Additionally management communication initiatives are implemented, where strategic issues are cascaded from the top of the company to the workshop level including a feedback loop. These exercises are obviously contributing to the creation of a European identity.

1.6.5.4 European identity by interactions between individuals

Another interesting approach when looking at European identity can be derived from the interactive sociology of the late 1950s: identity being the result of individuals constituting themselves as groups on the basis of interactions among themselves.[196] This is to be seen mainly in the context of the creation of nation states in the 19th century, criteria being e.g. founding fathers, common history, heroes, language, cultural assets, and traditions. With only 10 years of existence, its history, its different cultures and traditions, a European Astrium identity is of course still a way ahead. Nevertheless, language is an interesting issue: in the 18th century, the entire European elite spoke French as an instrument for the integration of nations.[197] Today the languages in Europe are rather an instrument for separation. Notwithstanding the weakness of the argument, Astrium's company language is English, which is as challenging for the French and German employees as it is a factor of integration.

1.6.5.5 European identity by polycentric and flexible patterns

On the basis of these deliberations, a promising and future oriented approach is elaborated: European identity should look for new topologies of the post-modern age: polycentric, flexible, with open patterns of relations, on the basis of the new European leitmotiv: "unity in diversity". In this sense, European identity is a structure that allows for coping with today's challenges and preparing for the future by using new resources.[198] Applying this new approach to Astrium shows great potential: Astrium has several polycentric poles of activities and competences; Astrium has flexible structures with the business unit and national representative organisational principles; Astrium allows for open patterns of relations between employees and organisations as they have to cooperate even though they have been strong competitors in the past. When understanding the creation of European identity as an open process with these attributes, Astrium has considerable potential to create a European identity in space.

1.6.5.6 European identity and time

European identity is also dependent on time. When comparing the 1973 to 2005 public EU polls, the index for European identity doubled on average during this period, and the index for European identity increased considerably after introduction of the Schengen agreement and the Euro.[199] The same goes for news on integration and symbols. Transferring this approach to Astrium, the factor of time obviously will count for Astrium. The extremely active and professional Astrium internal communication function providing information is a positive identity building factor. Additionally, there are plenty of articles and news on radio and TV covering Astrium and its activities. Projects in Astrium as symbols are important to this common identity. There is nothing more powerful in creating a European (Astrium) identity than the launch of an Ariane 5 launcher under the prime contractorship of Astrium or the start of operations of an Astrium built satellite (such as GOCE, Herschel and the telecommunication satellites in 2009). Summarising, Astrium has significant potential in creating European identity when looking at the criteria of time, news and symbols.

1.6.5.7 European identity by "Doing It"

There is another extremely interesting aspect to European identity: debates and controversial discussions can contribute to acquiring a European identity.[200] The same approach is applied when saying that national public debates have a key function in creating, deconstructing and reconstructing European identity.[201] Having been a stakeholder in European space activities for less than 10 years, being a truly European company based in five countries with a long history of individual companies that were heavily competing before joining in Astrium, there is substance for confrontation and controversial discussions. Media and competitors highlight the Astrium frictions and challenges, governments have been known not to behave neutrally in certain circumstances, and the internal debates among the Astrium employees and management have to cope with these issues as well. But – and this is an extremely encouraging aspect – all these debates help create a European and at the same time an Astrium identity. What counts is "doing Europe", i.e. negotiating and implementing integrated values in European "every day life"[202]: Astrium fulfils this requirement.

1.6.5.8 European identity by being an international actor

Another aspect of how to identify European identity is the international behaviour of an entity and its character as an international actor.[203] Astrium acts as international partner on several levels: (1) Astrium is the industrial main contractor to ESA for many programmes and projects.[204] This prime contractorship includes all relations to ESA as customer and all relations to the various suppliers. (2) Astrium has many bilateral industrial co-operation agreements and relations with partners throughout the world.[205] (3) With its subsidiaries in the service business, Astrium acts worldwide as a provider of communication services (mainly for UK, German, and French armed forces), Earth observation based services (Spot Image and Infoterra) and navigation with several pilot projects. (4) Finally, Astrium deals with several competing space companies on a national, European and international level, be it on the commercial market or on institutional levels. Obviously this is an extremely strong factor in creating a European Astrium identity.

1.6.6 Conclusion

It can be stated that there exists a considerable number of criteria for identifying and postulating a European identity. When applying these criteria to Europe's integrated space company, Astrium, it becomes clear that Astrium contributes to the creation of a European identity. The main aspects are:

1. The power of Astrium (programmes, economic key figures);
2. The Astrium corporate identity and its daily implementation;
3. Consciousness of having a European identity and accepting this;
4. Participative activities by management;
5. A structure that allows for coping with challenges and preparation of the future by using new resources;
6. Time of integration, news, and symbols of identification;
7. Confrontation, debates and controversial discussions contribute to acquiring a European identity;
8. Behaviour and character as an international actor.

A rather promising aspect is that the creation of a European identity is a dynamic, open process which is not closed or limited. What counts is "doing Europe" and that Astrium as Europe's only truly integrated company contributes substantially to the creation of a European identity.

Summarising it can be stated that Astrium as an integrated European company has generated over the first decade of its existence a process of creating a European identity, at the same time maintaining national and even regional identities. This is almost a synthesised dialectic of identities. Hence it merits that we conclude by quoting Claudio Magris in his work "Danubio" in 1986: "To acquire a new identity does not mean to abandon the first one, it rather enriches the own personality with a 'supplément d'âme'".[206]

[162] Nieder, Barbara. „Europäische Identität und Schule: Wie kann europäisches Bewusstsein gefördert werden?" Europäische Identität als Projekt. Thomas Meyer and Johanne Eisenberg (eds.) Wiesbaden: VS Verlag für Sozialwissenschaften 2009: 129.

[163] Cerutti, Furio. „Warum sind in der Europäischen Union politische Identität und Legitimität wichtig?" Europäische Identität als Projekt. Thomas Meyer and Johanne Eisenberg (eds.) Wiesbaden: VS Verlag für Sozialwissenschaften 2009: 250.

[164] Holfelder, Christian. „Ausgewählte Definitionen des Begriffs Corporate Identity." 20 October 2009. Corporate Identity Management. 13 October 2010. *http://www.corporate-identity-management.de/seite-6.html.*

[165] Eurospace, European Space Directory (ed.) Paris: Eurospace 2008: 22ff.

[166]ESA is an international organisation with 18 European Member States and has been Europe's space agency for more than 35 years. Today ESA disposes of a budget of some 3,5 billion euros per year, financed by ESA's Member States.
[167]European Space Agency. Convention for the Establishment of a European Space Agency, Article II. Doc. CSE CS(73)19, rev. 7 of 30 May 1975. Paris: ESA.
[168]*Ibid.*
[169]For further insights into the generation of European identity by the institutional stakeholders in Europe see the contributions of Jean-Jacques Dordain, Stephan Lingner, Alan Beward, Nina-Louisa Remuß in this book.
[170]The German newspaper FAZ put this on the point a few years ago by stating that Ariane has generated European identity. FAZ Paris correspondent Gerald Braunberger in FAZ of 21 May 2001.
[171]Spude, Mathias. Europe's new system Global Monitoring for Environment and security – benefits for society and industry at the same time. IAC-07-E3.3.01. International Astronautical Congress, 24–28 September 2007, Hyderabad, India.
[172]Schrogl, Kai-Uwe and Rohner, Nicola. „Für einen neuen Ansatz zur Begründung der Raumfahrt." Die Zukunft der Raumfahrt, ihr Nutzen ihr Wert. Carl Friedrich Gethmann, Nicola Rohner and Kai-Uwe Schrogl (eds.) Graue Reihe Nr. 40. Bad Neuenahr-Ahrweiler: Europäische Akademie 2007: 139ff.
[173]Eurospace, European Space Directory (ed.) Paris: Eurospace 2009: 72ff.
[174]Spude, Mathias. „Integrierte Zusammenarbeit, Die Europäische Weltraumorganisation (ESA)." Handbuch des Weltraumrechts. Karl-Heinz Böckstiegel (ed.) Köln: Carl Heymanns Verlag 1991: 667 ff.
[175]See the work of the German public law professor Georg Jellinek.
[176]Astrium website. 13 October 2010. *www.astrium.eads.net.*
[177]For a history of the creation of EADS see: Gadault, Thierry and Bruno Lancesseur. Jean-Luc Lagardère, Corsaire de la République. Paris: Le Cherche Midi, 2002.
[178]For an overview see the Astrium website at: www.astrium.eads.net.
[179]See, e.g.: Birkigt, Klaus, Stadler, Marinus and Hans J. Funck. Corporate Identity. München: Redline Wirtschaft 2002.; Holfelder, Christian. „Ausgewählte Definitionen des Begriffs Corporate Identity." 20 October 2009. Corporate Identity Management. 13 October 2010. *http://www.corporate-identity-management.de/seite-6.html.*; Fachhochschule Stuttgart/Hochschule der Medien. „Corporate Identity und Corporate Design." Website PR. 13 October 2010. *http://v.hdm-stuttgart.de/projekte/websitepr/cicd.php.*
[180]Birkigt, Klaus, Stadler, Marinus and Hans J. Funck. Corporate Identity. München: Redline Wirtschaft 2002: 18.
[181]Astrium. Astrium Corporate Management Principles and Responsibilities, Astrium Blue Book. December 2007. Chapter 2: Missions of Astrium Entities (not public).
[182]*Ibid.*, Chapter 1.1: Astrium Goals (not public).
[183]See the end of Chap. 2 of this article.
[184]Astrium. Astrium Corporate Management Principles and Responsibilities, Astrium Blue Book. December 2007. Chapter 1.1: Astrium Values (not public).
[185]*Ibid.*, Chapter 2: Mission of Astrium Entities (not public).
[186]For example, Astrium Satellites – being Europe's main industrial prime contractor for telecommunications, Earth observation, scientific and navigation satellites – has sites in Toulouse and Elancourt in France, in Ottobrunn and Friedrichshafen in Germany, in Stevenage and Portsmouth in the UK, in Madrid in Spain and in Leiden in the Netherlands. Astrium Satellites is managed by a CEO and an Executive Committee composed of managers from several countries with transversal „across the borders" responsibilities.
[187]See Sect. 2.2 of this article.
[188]As regards the factor of time in the creation of identities, see Chap. 5.6 of this article.
[189]Fachhochschule Stuttgart/Hochschule der Medien. „Corporate Identity und Corporate Design." Website PR. 13 October 2010. *http://v.hdm-stuttgart.de/projekte/websitepr/cicd.php.*

[190]Meyer, Lena. Deutsch-französische Unternehmenskultur am Beispiel der EADS Space Transportation GmbH. Diplomarbeit and der FHWT Vechta, Diepholz, Oldenburg 2006.
[191]Astrium. Astrium Corporate Management Principles and Responsibilities, Astrium Blue Book. Dec. 2007. Chapter 4.16.a: Communication & Public Relations Mission Statement (not public).
[192]This is valid for the internal audience who are addressed by "Flash-In" email-information, the internal company magazine, and dedicated events. The external audience is addressed via media relations (almost 2 press actions per week resulting in Astrium being the most mentioned space company in the media in Europe), via the Astrium website, via events (from events for dedicated decision makers to grand public activities, e.g. an Ariane 5 mock-up on Champs-Elysées), via exhibitions and trade shows, via visitor service at the Astrium sites, brochures, advertisements, etc.
[193]See, e.g. Kunczik, Michael. Public Relations, Konzepte und Theorien. Köln: Böhlau 2002: 22ff, 160ff., 338f.; Haywood, Roger. All about Public Relations. Cambridge McGraw 1991: 13ff.; Jefkins, Frank. Public Relations Techniques. Oxford: Butterworth-Heinemann 1994: 304ff.
[194]Meyer, Thomas. „Europäische Identität." Europäische Identität als Projekt. Thomas Meyer and Johanne Eisenberg (eds.) Wiesbaden: VS Verlag für Sozialwissenschaften 2009: 20ff.
[195]*Ibid.*: 23f.
[196]See Thiesse, Anne-Marie. „Die Europäische Identität: Erbe der Vergangenheit oder Konstruktion für die Zukunft?" Europäische Identität als Projekt. Thomas Meyer and Johanne Eisenberg (eds.) Wiesbaden: VS Verlag für Sozialwissenschaften 2009: 32ff.
[197]*Ibid.*: 42ff.
[198]*Ibid.*: 45f.
[199]Bruter, Micheal. „Institutionen und die europäische Identität der Bürger." Europäische Identität als Projekt. Thomas Meyer and Johanne Eisenberg (eds.) Wiesbaden: VS Verlag für Sozialwissenschaften 2009: 53ff.
[200]Glasson Deschaumes, Ghislaine. "Europa als Grammatik." Europäische Identität als Projekt. Thomas Meyer and Johanne Eisenberg (eds.) Wiesbaden: VS Verlag für Sozialwissenschaften 2009: 74.
[201]Liebert, Ulrike. „Ist eine europäische Identität notwendig und möglich?" Europäische Identität als Projekt. Thomas Meyer and Johanne Eisenberg (eds.) Wiesbaden: VS Verlag für Sozialwissenschaften 2009: 90ff.
[202]*Ibid.*: 108.
[203]Télo, Mario. „Die internationale Dimension der europäischen Identität." Europäische Identität als Projekt. Thomas Meyer and Johanne Eisenberg (eds.) Wiesbaden: VS Verlag für Sozialwissenschaften 2009: 169.
[204]Such as Ariane 5, Columbus laboratory and ATV for the International Space Station, several Earth observation satellites (e.g., ENVISAT, GOCE, CRYOSAT, EARTHCARE), scientific missions (e.g. Herschel/Planck, GAIA, Bepi Colombo) and navigation satellites.
[205]Such as with Russian Khrunichev company concerning the Rockot small launch vehicle, with Kazakhstan in developing Earth observation satellites including ground segment, with South Korea for Korean COMS (communication and meteorology mission), with NASA concerning the operations of a cargo carrier for the U.S. Shuttle and many others.
[206]Quoted in: Nieder, Babette. „Wie kann europäisches Bewusstsein gefördert werden?" Europäische Identität als Projekt. Thomas Meyer and Johanne Eisenberg (eds.) Wiesbaden: VS Verlag für Sozialwissenschaften 2009: 129ff.

Applications

1.7 Europe's relations with the wider world – a unique view from space

Alan Belward

1.7.1 Introduction

Europe is not the largest continent – in fact at 6,8% of Earth's land area it is the second smallest. Nor is it the most densely populated; the 50 states residing on this landmass are home to around 11% of Earth's population, making it the third most densely populated continent. No geopolitical entity groups all 50 countries, no common geophysical characteristics exist– diversity is a prominent feature of Europe.

Nevertheless Europe is a major player on the world stage. The 27 Member States comprising the European Union (EU) are the biggest aid donor in the world, consistently accounting for more than half of all Official Development Assistance in any year;[207] the EU is the largest trading block in the world, accounting for one fifth of global trade;[208] and the EU plays an increasingly prominent role on other global issues, especially environment, conflict prevention and crisis management. This list underscores two contrasting aspects of the emerging European identity – on the one hand "benefactor" and on the other "competitor".

The Treaties of Rome (at the root of today's EU) were signed in 1957, coincidentally the same year the Soviet Union launched Earth's first artificial satellite into space. The unique view of our planet from space increasingly informs relations within and outside the EU's borders and helps confirm that duality in Europe's identity does not equate to conflict in identity. Indeed the view from space emphasises the synergy between Europe as an economic power and a benefactor keen to promote transparent governance, responsible environmental behaviour and equity among peoples.

1.7.2 It's a small world

Seen from space Earth appears a harmonious whole; the continents, oceans and clouds seamlessly connect to form the familiar 'blue marble'.[209] Closer to home the situation is less harmonious, less whole, but it is fundamentally connected. European policy increasingly reflects this connectivity; our competitive identity is not viewed in isolation from global responsibilities. This interdependence is clear for example in Europe's approach to creating 'a resource-efficient Europe'.[210] This policy explicitly links economic growth and job creation in Europe with the global dimensions of issues such as climate change, land use, deforestation, and securing reliable access to supplies of natural resources. An example from

Fig. 1.14. *High-resolution imagery of the Congo Basin forest. The image is from the panchromatic imager on SPOT 5 (source and copyright CNES, processed image JRC). This illustrates how high-resolution optical data can be used for the surveillance of forests. The image shows a very small patch of the Congo forest (about 4 by 4 km) but shows it in immense detail. The white cross dividing the image into four quadrants is part of a network of logging roads crisscrossing the forest. The top and bottom right quadrants contain undisturbed tropical forest. The resolving power of the sensor (2.5 m) that acquired the image is fine enough to see each individual tree's canopy as a separate entity. The white "holes" seen in the top and bottom left quadrants occur where bare earth has been revealed through the felling of individual trees. Civilian satellites capable of even more detailed imaging than this now fly. These can resolve down to 41 cm, and at this resolution it is not individual tree canopies that are seen so much as individual branches in the canopy – at least when the leaves do not obscure the view.*

the paper illustrates; "land used to produce food may compete with land use for energy and both may compete with land which supports biodiversity or provides ecosystem services such as absorbing carbon from the atmosphere." – i.e., our need to import and export foodstuffs must be balanced with responsibility to the human population at large and to the natural world.

Policy goals such as a resource-efficient Europe reinforce the idea that prosperity, competitiveness and growth are achievable alongside sustainable development and poverty reduction in developing countries. This implies an ability to see Europe as an entity in its own right (the EU market, the EU energy supply...) and a part of the global community. Looking beyond our borders has long been a facet of European identity – European seafarers from the 10th century Norse sailors onwards and the land journeys of Marco Polo immediately spring to mind. To these viewpoints we must now add those from spacefaring.

1.7.3 Europe's spacefaring capabilities

Photography and manned flight developed almost contemporaneously in the early 19th Century. By 1858 Gaspard-Félix Tournachon – better known as

Fig. 1.15. *Logging roads and environs in the Congo Basin forest imaged by SPOT 20 m resolution multispectral imager (source and copyright original data CNES, processed image JRC). The black rectangle shows the location of Fig. 1.14. Images like this are being used to measure rates of deforestation throughout the tropics. Deforestation in the Congo Basin currently occurs far more slowly than in S.E. Asia or Latin America.*

"Nadar" – had combined the two technologies and was publicising the strategic and cadastral applications of photography taken from balloons. The desire for photographic information from above led to other innovations, such as the pigeon-borne cameras flown by the Bavarian Pigeon Corps,[211] though aircraft soon replaced this short-lived innovation. Cameras were among the first instruments put on Earth orbiting satellites and the first pictures of Earth from these new vantage points were obtained in 1960. Satellites now carry sensors far more powerful than the photographic and television cameras of 50 years ago. There are imagers capable of recording multispectral information beyond the range of the human eye's sensitivity, there are thermal sensors and there are imaging radar that can see through clouds and at night. Because these instruments are mounted on Earth orbiting platforms they can acquire imagery of anywhere on the planet's surface and they can do so again and again. Some satellites gather imagery every

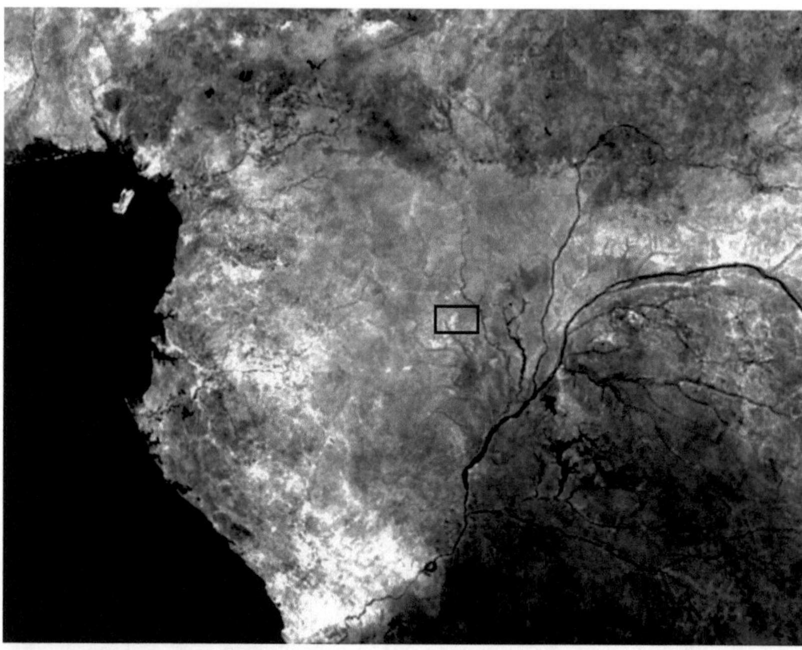

Fig. 1.16. *Congo Forest Basin imaged by the Vegetation sensor on SPOT 4 (source and copyright original data CNES, processed image JRC). This is a system capable of producing images at this scale every day. For many years satellite imagery of this type along with meteorological information and agro-meteorological models have been used for crop yield monitoring in Europe. These techniques are now being adapted for countries vulnerable to food crises and shortages. The system generates monthly reports on crop condition, estimates yield prospects and identifies the likelihood of food shortages and the European team works in close co-ordination with the UN World Food Programme and the FAO in this domain. The black rectangle shows the location of Fig. 1.15.*

15 minutes – though the smallest objects detected are measured in kilometres and the view is restricted to one side of the planet only (from a geostationary orbit). Others collect data a few times a year but such that objects less than 0.5 metres in size can be seen, no small achievement from over 600 km in space. (Figs. 1.14–1.17 provide examples of the different scales of imagery available today.)

The first European satellite with imaging capabilities was launched in 1977. This has been followed by at least thirty imaging satellites to date. These include some notable "firsts" – the first stereographic capability from space, first advanced Radar imaging, first microsatellites with an imaging capability, first thermal systems on microsatellites and the first simultaneous launch of an entire constellation of imaging platforms (see Table 1.4).

The list of platforms in Table 1.4 is testament to Europe's scientific and technical capabilities in the Earth observation domain. But European countries

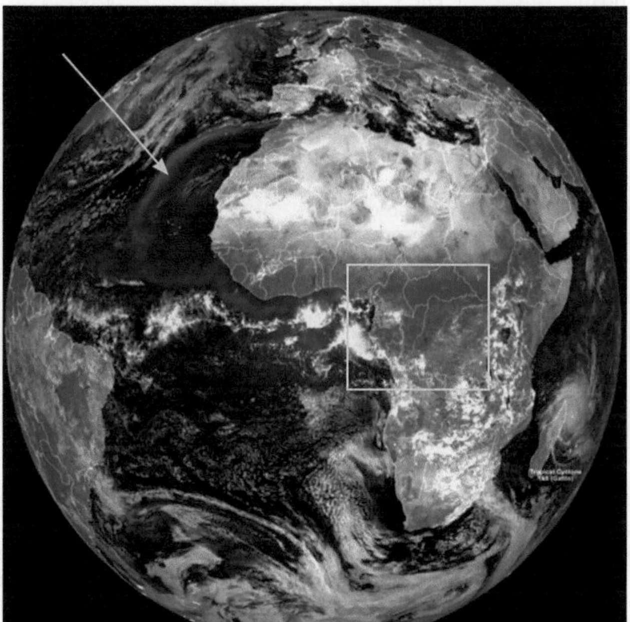

Fig. 1.17. *Meteosat image (original data copyright EUMETSAT 2005, additional processing JRC). Images like this are generated every 15 minutes and are a cornerstone of operational weather forecasting. The data are increasingly used in other environmental monitoring programmes too. The grey rectangle shows the location of Figure 1.16. The grey arrow highlights a plume of dust blowing off the Sahara Desert. Dust storms from the Sahara are on the increase, possibly as a consequence of climate change, but exacerbated by land clearance and over-cropping of marginal lands. This particular plume was transported across the Atlantic Ocean and deposited in the sea off the Florida coast, where it enriched the waters and induced a toxic bloom of considerable size – a graphic example of our connected world.*

Table 1.4: *History of European satellite Earth observing systems*

Date	Event
23rd November 1977	Meteosat 1 – Europe's first satellite with imaging capabilities supports operational meteorological services
19th June 1981	Meteosat 2
22nd February 1986	SPOT 1 – first stereoscopic imagery available, first 10m resolution imagery (in service till 31st December 1990)
15th June 1988	Meteosat 3
6th March 1989	Meteosat 4
22nd January 1990	SPOT 2 (de orbited 30th June 2009)
2nd March 1991	Meteosat 5
17th July 1991	ERS-1 – first C band space borne synthetic aperture radar (SAR) (ERS-1 in service till 10th March 2000)
26th September 1993	SPOT 3 in service till 14th November 1996
20th November 1993	Meteosat 6
21st April 1995	ERS-2 (in service till 22nd June 2003)
2nd September 1997	Meteosat 7
24th March 1998	SPOT 4
22nd October 2001	BIRD (Bispectral InfraRed Detection) the first microsatellite with thermal imaging capability
1st March 2002	ENVISAT the world's largest environmental satellite, carries 9 instruments (including optical and radar imagers)
4th May 2002	SPOT 5 – offering 2.5 and 5 m pan and 10 m MSS
28th August 2002	Meteosat 8 first of the Second Generation systems with 12 channels, + enhanced spatial and temporal resolution
28th November 2002	DMC – Europe's DMC launches its first operational microsatellite with imaging capabilities (for Algeria)
27th September 2003	DMC – launch of UKDMC1 (a 30 m MSS system) along with satellites for Nigeria and Turkey
21st December 2005	Meteosat 9
27th October 2005	TopSat launched – optical imager transmitting directly to a mobile ground station, provides rapid information delivery
19th October 2006	MetOp-A the first European polar-orbiting meteorological satellite

7th June 2007	COSMOSkyMed the first of a four craft constellation with X band Synthetic Aperture Radar
15th June 2007	TerraSAR-X variable resolution X band SAR offering 1 m to 16 m resolution imagery
9th December 2007	COSMOSkyMed 2 launched
29th August 2008	RapidEye 1 to 5 the first simultaneous launch of a constellation of high resolution (6.5 m) MSS imaging systems
24th October 2008	COSMOSkyMed 3 launched
29th July 2009	UKDMC2 and Demios 1 offer improved calibration and improved spatial resolution over the first generation DMCs
2nd November 2009	SMOS soil moisture and ocean salinity mission launched
8th April 2010	CryoSat 2 carries a SAR Interferometric Radar Altimeter
5th November 2010	COSMOSkyMed 4 completes the constellation of X band SAR carrying platforms

are not the only spacefaring nations in the world. The Committee on Earth Observation Satellites (CEOS), established by the G7 twenty-five years ago brings together most governmental organisations responsible for civil space-borne Earth observation activities. Collectively these agencies operate over 100 Earth-observing satellites.[212]

1.7.4 Policies and applications

An overview of policy areas supported through Earth observation science and technology is shown in Table 1.5. Collectively they give an idea of just how broad Europe's policy interventions are on the world stage. The hundred or so satellites flying provide data with which to support these diverse policies.

Policy support from Earth observation takes many forms: intelligence gathering through surveillance and mapping operations, inventories through mapping and measurements, forecasts and scenario building through modelling. One common factor is the unifying effect of viewing our planet from space – images from Agadez to Aberdeen, Zeebrugge to Zanzibar are obtained with exactly the same accuracy, the same precision and the same information content. Space-borne imaging systems have the ability to locate detailed information for practically any point on the planet's surface, and this information can be delivered within hours, if not minutes of the satellite passing overhead. Such wealth of information brings new

Table 1.5: *Policies and their related application areas where considerable use is made of EO to support the EU's international relations*

Policies	Agriculture, Development, Energy, Environment, Foreign and Security, Humanitarian Aid, Maritime, Neighbourhood, Research, Space, Trade					
Policy Action Areas	Agricultural production	Marine and Maritime	Crisis Management	Natural Hazards	Resources Management	Climate Change
Applications	Land use and subsidy controls	Vessel detection and fishing area mapping	Crisis anticipation and prevention	Fire forecasts and impact assessments	Water availability and quality monitoring	Observations, assimilation and modelling
	Yield estimations	Pollution detection and control	Post-crisis recov-ery and reconstruction	Flood forecasting and impacts	Forestry and land cover monitoring	Monitoring, reporting and verification
	Food security assessments	Environmental indicators	Risk reduction	Drought monitoring	Biodiversity and protected areas	Reducing scientific uncertainty
	Land suitability and availability assessments	Surveillance (piracy and migration)	Emergency mapping	Desertification mapping and monitoring	Soils and minerals	Calibration, validation and benchmarking

dimensions of transparency, accountability and even "policing" to international relations.

For example, the metal hulls and superstructure of shipping make their detection, and even sometimes identification (i.e., putting a name to a ship) possible from space-borne imaging radars. Space-based vessel detection systems support applications such as fisheries control by helping to maintain quotas and ensuring that the location and composition of fishing fleets corresponds with legal agreements between countries – the satellites act as an international referee, assuring fair play and providing transparency between Europe and her fishing neighbours around the world.

It is a short step from refereeing to policing and this role too is partially met through satellite surveillance. This may be benevolent, such as helping in international ship routing and navigation or may have more of an enforcement character. In addition to being good for detecting ships, radar data are also excellent for measuring oil-spills on the surface of the sea. This combination offers great potential for the control of illegal oil dumping, or the flushing of oil tanks out at sea. Some of the very high-resolution radar systems now available are also being tested as part of Europe's response to smuggling, immigration, piracy and other safety and security threats.

Surveillance applies equally well to the natural environment. Technology that can detect a single ship or map individual buildings can also be used to monitor the felling of individual trees. Because the images are obtained from satellite this capacity is not restricted to the EU's territory but can occur anywhere, even in places as remote as the middle of the Congo forest. At first glance having such a capacity doesn't appear to have much to do with Europe's international relations. But this technology provides confirmation that the forest is being exploited responsibly and sustainably. Such systems can also help confirm whether logging is legal or not (e.g., inside or outside the boundaries of a forest reserve). If the technology is transferred into local hands (as is the case in fact with the Congo) then it not only builds scientific capacity in the recipient country it also provides that country with the capacity to police and to manage its own resources – contributing to Europe's position as global leader in development assistance and providing concrete benefits towards poverty alleviation. Finally it offers considerable potential as a tool for monitoring, reporting and verification work in the context of international conventions, such as the recent discussions in the UN Framework Convention on Climate Change (UNFCCC) on Reduced Emissions from Deforestation and Degradation[213] and thus benefits the global community; helping design and test such tools is of course an explicit European contribution to this global community, but it is also true that the knowledge gained, and the knowledge disseminated add to our own negotiating power and consequently reinforce our identity as a key player on the global 'environment' stage.

The UNFCCC is one of the defining global policy areas of the early 21st Century. Climate change is a truly global phenomenon, and one that calls for global action. The global climate observing system has identified 44 sets of observations – referred to as Essential Climate Variables (ECVs) – as the minimum needed to provide reliable climate information[214] and at least 31 of these ECVs can be measured from space.[215] The systematic observation of climate variables is key to improving our ability to model current climates and predict change. And Europe is playing an important role in meeting these commitments e.g. monitoring environmental factors such as forest and savannah fire occurrence[216] and for the creation of measurements of surface albedo.[217] The possession of satellites and the ability to harness the data they provide reinforces our sense of Europe as "one of the good guys'", contributing to global science, contributing to understanding, contributing to the common good.

1.7.5 A framework for sustained services

Observing our planet from space calls for advanced industrial and engineering skills. Europe is able to design, build, launch and fly imaging satellites, retrieve the data they produce, analyse it and put it to good use. This capacity reinforces Europe's identity as a world-class industrial power, as an innovator and manufacturer.

Europe has taken the lead in small satellite technologies[218] for example. Results of this are innovative new approaches to Earth observation (why launch one imaging satellite when you can launch five at once?) and building new overseas markets (imaging small-sats have been built for Algeria, Nigeria, China and Turkey as well as European countries). Small-sat constellations, such as RapidEye and DMC (see Table 1.4) offer certain advantages over their bigger 'traditional' single platform rivals. Multiple satellites working together allow rapid access to data, provide more regular imaging of any given point on the Earth's surface, offer more opportunities to see through gaps in the clouds and thus give more frequent updates of any changing situation (from the aftermath of catastrophic events such as floods, fires and earth movements to the steady growth of crops through a season and the changes in our planet's forest cover over years).

Whilst Europe is making significant advances at the level of the small-sat 'entrepreneur' it is also progressing at the level of pan-European government sponsored activities. The European Commission and European Space Agency have established the Global Monitoring for Environment and Security (GMES) programme to provide a framework for sustained operations and therefore also long-term support for Europe's space industry.[219]

The European Union's Earth observation programme (GMES) entered into force on 9th November 2010. This programme includes building and flying some of the most technologically advanced imaging satellites in the world (the Sentinels) as well as running services in domains such as emergency response, land, ocean and atmosphere monitoring and eventually full climate and climate change monitoring services.[220]

Promoting innovation and economic growth in Europe's aerospace industry are explicit goals of the GMES programme. This role as a policy instrument to advance industrial growth is balanced by the operational focus of the programme. Information from GMES is designed to help policy-makers prepare national, European and international legislation on environmental matters including climate change, for cooperation linked to development, humanitarian aid and emergency situations worldwide and to address security matters (e.g. border surveillance). The programme both informs the preparation of such policies and helps monitor their implementation. Again a European identity embracing both global vision and more parochial concerns becomes apparent.

1.7.6 Conclusions

Monitoring international situations and conditions, building capacity for the benefit of others, encouraging good governance and aspirations to global environmental stewardship are recurrent features of Europe's international relations – all are helped and honed by our capacity for Earth observations on the global scale.

Europe's space policy and the GMES programme which forms part of it, provide a clear European framework for Earth observation applications – a framework that aims to satisfy public interests, support economic growth and benefit society. Earth observation science and technology has already become an indispensible part of many public services in Europe and GMES will ensure these continue well into the future. These services provide consistent and verifiable information on conditions across the territory of the EU, for neighbouring European countries and indeed the world.

The information Europe gathers through surveillance, the maps and measurements we produce along with the outcomes of the models we run certainly reflect European philanthropy. We seek to advance our understanding of the way in which the planet functions for the good of all, not just for the good of Europeans.

But our international relations are not exclusively altruistic, and the view from space provides us with tangible strategic and commercial benefits. We can track the location and condition of global resources, we can anticipate threats to our territory and we also sell satellites and the data they generate in the global marketplace.

The unique view from space is increasingly informing both policy making and policy implementation. Policies promoting European industrial excellence, innovation and creativity with policies that protect the environment, combat poverty and enhance the quality of life benefit in equal measure. As a result a European identity whereby the economic wellbeing of a successful competitor is not in conflict with the good behaviour of a benefactor becomes a reality. The Lisbon Treaty, which entered into force on 1st December 2009, calls for a European space policy to promote scientific and technical progress, industrial competitiveness and the implementation of its policies,[221] which neatly captures the altruistic and competitive sides of European international relations. Whether driven by discovery, science or commerce it is certain that the view from space makes our planet appear smaller than ever – our neighbours are visible to us, and we to them. It also unequivocally reinforces the connections between us all – what we do affects our neighbours, what they do affects us, and yet the view from space provides us all with access to the same global picture.

1.7.7 Disclaimer and acknowledgments

The views expressed in this paper are purely those of the writer and may not in any circumstances be regarded as stating an official position of the European Commission. The applications identified throughout the paper result from the efforts of many scientists working at the European Commission's Joint Research Centre. Thanks also go to Katherine Jane Ridley for comments on the drafts.

[207] Donor Atlas 2008, 29 October 2010. *http://development.donoratlas.eu/.*

[208] "What is Europe's trade policy?" EC DG for Trade. 29 October 2010. *http://trade.ec.europa.eu/doclib/docs/2009/may/tradoc_143154.pdf.*

[209] View of the Earth as seen by the Apollo 17 crew travelling toward the Moon, 7th December 1972.

[210] European Commission. Communication from the Commission to the European Parliament, the Council, the European Economic and Social Committee and the Committee of the Regions. A resource-efficient Europe – Flagship initiative under the Europe 2020 Strategy. COM(2011) 21 final of 26 January 2011. Brussels: European Union.

[211] "History of Aerial Photography". Professional Aerial Photographers' Association. 29 October 2010. *http://www.papainternational.org/history.html.*

[212] CEOS. Committee on Earth Observation Satellites Earth Observation Handbook, Climate Change Special Edition 2008. 29 October 2010. *http://www.eohandbook.com/*.
[213] Achard F, Defries R, Pandey D, Souza C. "Overview of the Existing Forest Area Changes Monitoring Systems." Reducing Greenhouse Gas Emissions from Deforestation and Degradation in Developing Countries: a Sourcebook of Methods and Procedures for Monitoring, Measuring and Reporting. Eds. Achard F., Brown S., De Fries R., Grassi G., Herold M., Mollicone D., Pandey D., Souza C., Alberta: GOFC-GOLD Project Office, Natural Resources Canada, 2009: 134–139.
[214] GCOS. An Implementation Plan for the Global Observing Systems for Climate In Support Of The UNFCCC. WMO Technical Doc. WMO/TD 1219 of September 2004.
[215] GCOS. Systematic Observation Requirements for Satellite-based Products for Climate; Supplemental details to the satellite-based component of the "Implementation Plan for the Global Observing System for Climate in Support of the UNFCCC (GCOS-107). Doc. WMO/TD No. 1338.
[216] Roberts, G. and Wooster M.J. "Fire Detection and Fire Characterization over Africa using Meteosat SEVIRI, IEEE." Transactions on Geoscience and Remote Sensing 46 (2008): 1200–1218.
[217] Govaerts, Y., and Lattanzio, A. "Retrieval Error Estimation of Surface Albedo Derived from Geostationary Large Band Satellite Observations: Application to Meteosat-2 and -7 Data." Journal of Geophysical Research 112 (2007).
[218] Xue Y., Li Y. Guang J., Zhang X. and Guo X., "Small Satellite Remote Sensing and Applications – History, Current and Future." International Journal of Remote Sensing 29 (2008): 4339–4372.
[219] Commission of the European Communities. Communication. Global Monitoring for Environment and Security (GMES): We Care for a Safer Planet. COM (2008) 748 final of 12 November 2008. Brussels: European Union.
[220] Commission of the European Communities. Proposal for a Regulation of the European Parliament and of the Council on the European Earth observation programme (GMES) and its initial operations (2011–2013). COM(2009) 223 final of 20 May 2009, adopted on 9 November 2010.
[221] Council of the European Union. Resolution on "Taking forward the European Space Policy". Doc. 13569/08 of 29 September 2008. Brussels: European Union.

Security

1.8 Space and security as an identity forming element – meeting Europe's external and internal security through space applications

Nina-Louisa Remuss

Europe has long been characterised by war and conflict. Thus, the basic idea behind the European integration process was to establish peace and security through economic integration. In order to do so, a common understanding of "peace" and "security" needed to be laid down. Once this had been achieved the necessary instruments for providing security both internally and externally had to be found. One such instruments are space applications, which can in turn also contribute to the development of a European security identity. Thus this essay seeks to answer three questions: (1) How can "security" be an identity-forming element?; (2) How can space policy and particularly a European Space Security Policy contribute to a European identity? And (3) What specific elements of a European Space Security Identity can be identified?[222] This European Space Security Identity is one *tessera* in the process of establishing a European identity or one small piece of the already existing European identity.

Before however answering these questions, the basic concept of "identity" and the development of the "idea of Europe" in history will be outlined. On the basis of this, the different definitions of "security" will be shed light upon. In the following two case studies – the European Draft Code of Conduct and the development of a strategy for the provision of internal security – will serve as examples to identify the elements of a European space security identity. Some conclusions will then be drawn related to Europe's future steps. Finally, it will be highlighted what advantages such a development of a European space security identity bring for the citizens of Europe.

1.8.1 Background: the concept of "Identity" and the "Idea of Europe"

Europe is an historical idea with different connotations in different times.[223] The concept of identity has often been used as a synonym for unity among the people of Europe. The idea of Europe as a distinct unity is post-classical and was created in the Middle Ages. However, there was no single medieval conception of European unity but rather a number and not always compatible understandings of unity: people were unclear about what they thought of unity and which unity they desired. This was mainly due to the fact that during the Middle Ages there were two sources for the notions of a united society: (1) the idea of a Christian community and (2) the universalism of the Roman Empire. As a consequence people were thinking about Christian unity, which transcended European unity and was not a geographical idea at all and others confused religious and political unity. Additionally, it proved to be easier to create a sense of unity on the basis of hostility and opposition than in view of the positive benefits it might confer. Thus, it was only when Christian united against Islam that they were able to act together. The unity of Europe in the Middle Ages was a unity of civilisations but it was not a unity capable of taking political form.[224]

In the late 12th century European unity was achieved through repression and persecution of minorities, being essentially a result of violent homogenisation. With the fall of Constantinople in 1453 and the beginning of colonial expansion in 1492, the idea of Europe became linked to a system that was regarded as specifically European characterised by specific European values. When in 1492 America was discovered, Columbus became to replace Charlemagne as the symbolic figure that united Europe. The concept of the "continent" evolved, maps and books were invented and Europe was "realised". During this so-called "age of discovery" the question of European unity became part of the greater question of an international order with more and more countries turning their attention to countries overseas. Europe lost its association with Christendom and became understood as a geographical unity. With the decline of the Turkish supremacy after the Battle of Lepanto in 1571, the idea of Europe and the origins of the European identity came to be associated with the resistance to the Turks.

During the French Revolution (1789) and the rise of nationalism, Europe became associated with a purely geographical unity and national values came to replace European values as it was commonly believed that there was no higher unity than the nation-state. It was only after 1945 when the people of Europe lost their faith in nationalism that the European spirit was born.[225]

As it can be seen in this analysis of the historic development of European identity and unity, Europe is not only a geographical division but a cultural, ideological, political, etymological, philosophic, Christian, mythological, economic and/or security unity.

Benedict Anderson famously defined a nation as an "imagined community",[226] whose group identity is created through a shared past and tradition. This is done through a selective presentation of history by a movement that wants to distinguish itself from other unities and in order to do so aims at creating a sense of identity. In line with this and the above analysis, Europe can be defined through certain *lieux de memoires*, that are specific symbols of the nations' past such as Greek mythology, historical persons (like Charlemagne and Napoleon), peace treaties and political cooperation (the Congress of Vienna and Verdun), Antisemitism, the Crusades etc. Europe can also be defined against an "other".

1.8.2 Different definitions for "Security"

In today's post-Cold War security realm one can distinguish between external and internal security. "External Security" equals the classical perception of security and deals with territorial security where threats are mainly state-focussed coming from outside the state's boundaries. Those are mainly military threats, which are dealt with by the Common Security and Defence Policy (CSDP).[227] Space is already used as an instrument in the provision of external security.

"Security" has been redefined in the post-Cold War era. Moving away from the traditional perception of security, it is now being referred to as "functional security" which includes (non-military) threats coming from non-state actors even from within the own state's boundaries. Examples of EU policies dealing with the provision of internal security are Maritime Policy, Critical Infrastructure Protection Strategies, transportation security and counter-terrorism policy. Space as an instrument has only recently been acknowledged to contribute to these areas of the provision of internal security.[228]

Taking this into consideration, Europe needs to adopt, build and implement a *formative role* and *principled identity* in matters of space and security. This means that it needs to exert influence by normative action instead of just handling or administrating the status quo set by others (i.e., formative role) while at the same time ensuring coherence with fundamental European values as laid down in various documents such as the European Security Strategy (i.e., principled identity).[229]

1.8.3 Space and security

In the Europe of today, space applications, that is Earth Observation (EO), Satellite Navigation (SatNav) and Satellite communication (SatCom) are used to support the provision of both external and internal security.

1.8.3.1 Space applications supporting external security missions

Satellite imagery (EO) was for example used for mapping in support of the EUFOR Chad/RCA[230] or is currently used to identify pirates' skiffs and conduct risk mapping. Satellite imagery is also used by the European Union Satellite Centre (EUSC) to conduct imagery analysis. Combining this analysis with all other available data of the geographic information (e.g., collateral information, intelligence data etc.) the EUSC delivers products that can be easily used by planners, emergency responders and decision makers and provides a common operating picture (COP).

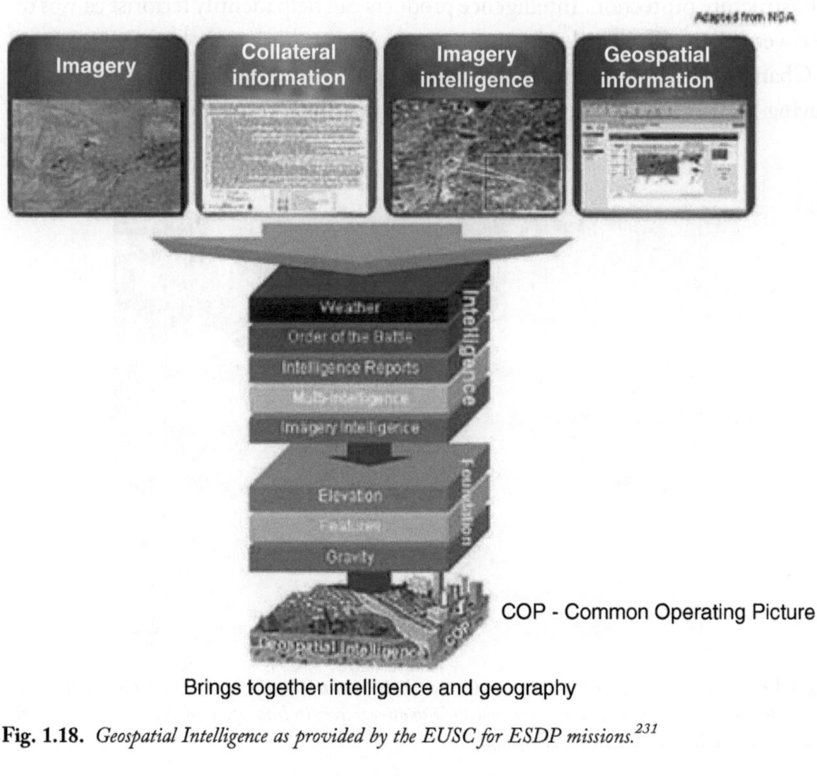

Fig. 1.18. *Geospatial Intelligence as provided by the EUSC for ESDP missions.*[231]

Satellite imagery can also be used to establish a data base of crisis prone areas. Stored data can eventually be compared with real-time satellite imagery. By conducting change detection, impact or damage assessment is possible.

SatCom proved indispensable in ESDP missions that were characterised primarily by the lack of supporting infrastructure in the host country and a particularly large area of operation. The great distances between the Operations Headquarters (OHQ) and field deployments can be overcome by using secure communication structures relying on SatCom. SatNav is primarily used for planning purposes but also for tracking of for example pirate skiffs, their mother-ships or hijacked ships.

1.8.3.2 Space applications supporting the provision of internal security

Satellite Imagery can also be used in the provision of internal security, i.e. in the fight against terrorism, in support of border management and as part of critical infrastructure protection. Intelligence products can help identify terrorist camps or discover ships with illegal migrants.

Change detection techniques can even be used to identify illegal migrants before leaving their mother country when they are boarding at the coast.

Fig. 1.19. *Picture shown to the U.N. Security Council during Colin Powell's 5 February 2003 presentation. Powell described it as showing a terrorist poison and explosive factory in Iraq, operated by an Islamic terrorist group, Ansar al-Islam, with ties to Al-Qaeda.*[232]

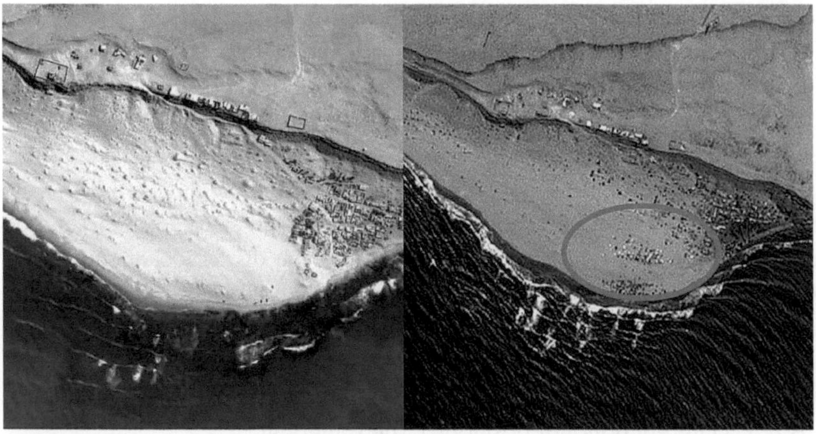

Fig. 1.20. *Coastal monitoring showing illegal migrants gathering for departure.*[233]

Satellite imagery can also be used to draw maps of critical infrastructures like oil refineries or nuclear power plants to conduct risk mapping and emergency planning.

SatNav and SatCom can be used to track cargo (e.g., transporting dangerous goods) and to establish risky transport routes for land, waterborne and air transport.

1.8.4 External security: the case of the Code of Conduct

In December 2008 the Council of the EU approved a draft Code of Conduct for outer space activities with the purpose to enhance the safety, security and predictability of outer space activities for all.[234] This European Draft Code of Conduct will serve as a case study to identify the elements of a European external security identity.

1.8.4.1 The setting

The existing legal situation provided for by the Outer Space Treaty of 1967, specific United Nations General Assembly Resolutions (e.g., the Resolution on the Peaceful Uses of Outer Space), UN treaties (e.g., The Registration Convention) and disarmament agreements (e.g., START) were no longer perceived as sufficient in the light of an increasing dependence on space assets. The two main institutions dealing with space and security are the Conference on Disarmament (CD) and the UN Committee on the Peaceful Uses of Outer Space (UN COPUOS). Since 1981 the UN has been passing an annual resolution calling for the continued peaceful uses of space and the prevention of an arms race in outer space (PAROS). Related discussions in the CD on the peaceful uses of outer space have been deadlocked since 1995 because of diverging national positions. China had bound the discussion of the Fissile Material Control Treaty (FMCT) to PAROS discussions, which proved unacceptable to the U.S.

1.8.4.2 Different national positions ("The Others"?)

In general it is possible to identify three possible routes to Space Security: The treaty approach, the Code of Conduct, i.e., a non-legally binding, approach and alternative measures such as Space Traffic Management (STM).

China and Russia favour a legally-binding treaty. They have been proposing a treaty on the peaceful uses of outer space since 1998. In this context China issued in 1998 the so-called Chinese Working Paper, which included a list of suggestions for ways to address PAROS. In 2002 Russia and China together tabled a joint-Russian-Chinese Proposal entitled Possible Elements for a Future International Legal Agreement on the Prevention of the Deployment of Weapons in Outer Space, the Threat of Use of Force Against Outer Space Objects. This was complemented by subsequent thematic working papers. In 2008 China and Russia tabled a third proposal entitled "Treaty on the Prevention of the Placement of Weapons in Outer Space, the Threat or Use of Force against Outer Space Objects" (PPWT).[235]

Such a treaty approach however was unacceptable for the U.S. under the Bush Administration, which was not open to discussions at all: "We are not prepared to negotiate on the so-called arms race in outer space. We just don't see that as a worthwhile enterprise." (John R. Bolton, Prev. Under Secretary of State for Arms Control and International Security).

1.8.4.3 The choice to be made

It is possible to identify a growing awareness of Space and Security in Europe. Space activities in Europe can be allocated to three levels: (1) ESA, EU (Council) and other intergovernmental programmes, (2) the European Commission

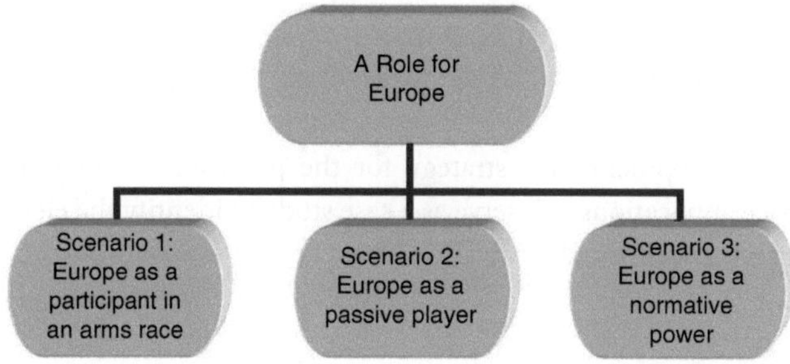

Fig. 1.21. *Europe's choice.*

(supranational) and (3) national space policies. In the context of the above-mentioned national positions, Europe needs to find its own role. While trying to find a role for Europe, three scenarios are thinkable: (1) Europe as a participant in an arms race, (2) Europe as a passive player or (3) Europe as a normative power adopting a formative role and principled identity. Europe's choice is depicted in the figure above.

1.8.4.4 Elements of a European Space Security Identity in the three scenarios

In all three scenarios different elements of a European Space Security Identity can be identified.

In Scenario 1, where Europe engages as an active participant in an arms race, it is possible to qualify this engagement as "balancing" against other powers. Europe would possibly introduce space-based and terrestrial weapons and no longer engage in international cooperation as this is perceived as too insecure. It would follow a purely military space policy in line with a hard power approach, which would include preparation for a military strike and deterrence.

In Scenario 2, where Europe decides not to decide, it would be a passive side player without a particular identity. As a consequence it would no longer be perceived as a credible international actor.

In Scenario 3, Europe is acting as a normative power or mediator. It would counter an arms race by influencing other actors. It would achieve change without having to engage in non-peaceful means in outer space. It would develop a particular space identity that would mainly include diplomatic tools and international treaties as its instruments. It would follow a soft power approach. This scenario is the closest to the proposed adoption of a formative role and principled identity.

1.8.5 Internal security: the case of the developing strategy for internal security

The possible development of a strategy for the provision of internal security including space applications will serve as a case study to identify the elements of a European internal security identity.

1.8.5.1 The setting

While the notion that space can play a key role in security policy is finally taking ground, its role in internal security is still neglected. Recently, some steps have been taken to increase research and development in the use of space applications for internal security. One can distinguish two levels of internal security activities in Europe: the European level (EU and ESA) and the national level. The fact that internal security is often neglected is particularly striking as the European Security Strategy (ESS) explicitly mentions terrorism and organised crime among the five key threats Europe needs to be prepared for. Thus, if space applications prove to be innovative and effective tools in the provision of internal security, why should research in this area not be increased? In June 2005 the Panel of Experts on Space and Security (SPASEC) was mandated to review the role of space in the fight against terrorism and thus did a first step in considering space applications for the provision of internal security. The subsequent SPASEC-Report, which was published in March 2005 explicitly considered the support of space-based applications in counter-terrorism operations. Three year long "Preparatory Action" (PASR) (2004–2007) addressed the "protection against terrorism" as one of its five main areas. The following European Security Research Programme (ESRP) of the FP7 includes several projects, which deal with space applications in the provision of internal security. Both, PASR and the FP 7 security theme were supported by two high-level strategy groups, i.e. the Group of Personalities (GoP) for Security Research and the European Security Research Advisory Board (ESRAB). The latter proposed the establishment of the European Security Research Innovation Forum (ESRIF). One of ESRIF's working groups is dealing specifically with "Situation Awareness and the role of space".

Generally speaking, Europeans feel less threatened by terrorism than Americans, i.e. 71% of Americans think it is likely that they will be personally affected by international terrorism while only 53% of Europeans feel this way and Europeans in turn feel more likely to be affected by global warming (73% Europeans vs. 64% Americans). On an even more general level, when asked about the importance of developing various space-based applications for Europe, EU respondents were keen on (further) developing environmental/natural monitoring systems with 58% finding this very important. Regarding other services the mood remained generally positive; the development of applications to monitor citizens' security was considered as at least quite important by 75%.[236] An overwhelming majority in most countries (most notably in e.g. Hungary, Romania, Ireland and Poland, about 80% in each) found the provision of space-based monitoring for security applications important. However, significant minorities do not consider so important or important at all this type of application in a number of countries. Those who believe that space-based monitoring that improve citizens security are not so important, or not important at all for Europe reached 46% in Austria, 43% in Germany, 39% in Sweden and 37% in Denmark and France.[237]

1.8.5.2 Existing documents and strategies

At EU level there currently exists a number of documents laying down EU's strategy in the fight against terrorism such as the Plan of Action on Combating Terrorism, the Declaration on Combating Terrorism, The Hague Programme and the European Union Counter Terrorism Strategy. The latter mainly provides for the four main principles in the fight against terrorism, i.e. prevent, protect, pursuit and respond. It also guarantees that the combat against terrorism at global level should respect human rights and make Europe safer, allowing its citizens to live in an area of freedom, security and justice. Additionally, there are various specific policies and strategies such as the European Neighbourhood Policy (ENP), The Maritime Strategy and Strategies related to organised crime and illegal immigration.

1.8.5.3 Elements of a European Space Security Identity in the provision of internal security

The EU started to look into the provision of internal security only recently as part of its FP 5 to FP 7. In discussions with the U.S. (e.g. Passenger Name Record (PNR)) one can identify an emphasis on freedom and justice. This is in line with the proposed adoption of a principled identity. There is currently no centralised European approach through a Department for Homeland Security or a related Directorate General (DG) in the provision of internal security in Europe. This leaves room for doubts about the existence of a European identity in this area. It is possible to identify a fragmentation among different policies (horizontal), different bodies and actors in the three pillars (institutional) and between the EU and its

Member States (vertical). This fragmentation resulted in incoherence and the absence of a holistic and integrated approach and hinders the development of an EU internal security identity.

1.8.6 Elements of a European Space Security Identity

A European Space Security Identity needs to feature and frame the existing principles and values derived from existing documents such as TEU Art. J.1., ESS, von Wogau Report etc., observable trends and official statements. It should account for new principles as derived from changing circumstances, modified assessments or from the evolving European integration process.

The fundamental objectives and principles of the European Space Security Identity should as part of the external security identity safeguard common values, preserve peace and strengthen international security, promote cooperation, democracy and the rule of law as well as multilateralism for an international order, follow a soft power approach through the combination of civilian and military tools and highlight the importance of diplomacy.

From the area of internal security, the European Space security should account for all internal security threats and not solely emphasise terrorism. In the provision of security it should highlight human rights and freedom and guarantee a strong protection of personal data. As a successful counter-terrorism policy it should rely on the Criminal Justice Model, fighting terrorism through prosecuting terrorists.

This European Space Security Identity is one piece of the European Identity.

Table 1.6: *Elements of a European Space Security Identity*

External Security	Internal Security
– Emphasis on multilateralism & cooperation *(formative)*	– Terrorism as one out of many internal security threats
– Promotion of democracy and the rule of law *(formative & principled)*	– Emphasis on human rights and freedom *(principled)*
– Combination of civilian and military tools *(formative)*	– Strong personal data protection *(principled)*
– Diplomacy (space as a tool) *(formative)*	– Criminal Justice Model as the basis for a successful counter-terrorism policy *(principled)*
	– Internal security in contrast to homeland security not only geographically defined

1.8.7 The way forward for Europe

Europe needs to develop a Space Security Identity, by adopting, building and implementing a *formative role* and *principled identity* in matters of space security. In the realm of external security, Europe needs to widen its concept of security, accounting for threats coming from non-state actors as well. It should establish a coordinated space dialogue with international partners and follow-up on Code of Conduct negotiations, thereby trying to get as many states as possible to adhere to the Code.

In the area of the provision of internal security, Europe needs to formulate a common European approach to internal security through an Internal Security Strategy show-casing a specific European approach or identity which stresses the importance of use of technology, particularly space, and provides for better coordination among projects and institutional actors, Member States, EU and ESA.

1.8.8 The European citizens

The majority of European Union citizens regard European space activities as important from the perspective of EU's future global role: one in five citizens considered such activities very important (20%) and a further 43% felt that space activities are important. Almost two-thirds of Europeans share the view that space activities are important for the future international position of the European Union.[238] Having identified the elements of a European Space Security Identity, the question arises what advantages such a development has for the European citizens. First of all they can be sure about a European approach stipulated by space as an instrument for providing security. To the inside, citizens would be assured that such an approach is in line with justice and freedom. To the outside they can be sure that it is promoting international cooperation, democracy and the rule of law. Through adopting a formative role and principled identity promoting certain specific European values, a group feeling is established: the citizens identify themselves with the European values and with Europe, as a result support for Europe increases and finally Europe increases its coherence and becomes a stronger actor.

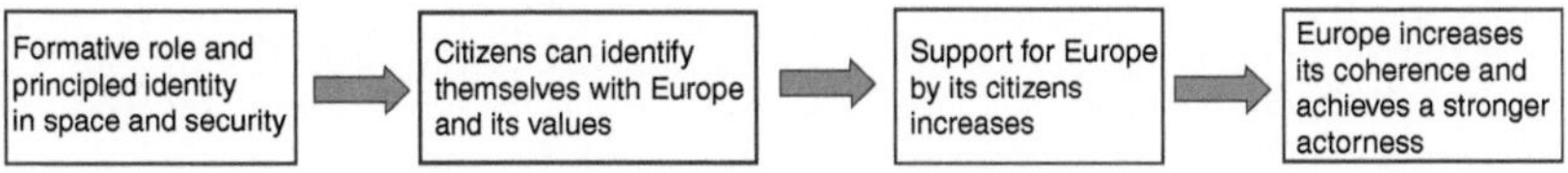

Fig. 1.22. *How the adoption of a formative role and principled identity contributes to Europe's actorness.*

[222]It is already well accepted that policies are dependant on Europe's general strategic direction. Yet, in how far policies themselves influence the strategic direction of Europe is rarely discussed.
[223]Barraclough, Geoffrey. European Unity in Thought and Action. Oxford: Basil Blackwell 1963: 3.
[224]*Ibid.*: 7, 22–5.
[225]*Ibid.*: 33–35, 46.
[226]cf. Anderson, Benedict. Imagined Communities: Reflections on the Origin and Spread of Nationalism. London: Verso (revised version) 1991.
[227]Previously referred to as European Security and Defence Policy (ESDP), the Lisbon Treaty now refers to the Common Security and Defence Policy (CSDP).
[228]For more information on research programmes dealing with space as an instrument in the provision of internal security see Remuss, Nina-Louisa. "Space and Internal Security – Developing a Concept for the Use of Space Assets to Assure a Secure Europe." ESPI Report 20. Vienna: European Space Policy Institute 2009: 11.
[229]Rathgeber, Wolfgang, and Nina-Louisa Remuss. "Space Security – A Formative Role and Principled Identity for Europe." ESPI Report 16. Vienna: European Space Policy Institute, 2009.
[230]Öller, Gustav. "European External Operations and Reliance on Space: A Case Study." Presentation. EC-ESA-EDA Workshop on Space for Security and Defence. Brussels, Belgium. 16 September 2009.
[231]"Geospatial Intelligence." European Union Satellite Centre. 22 March 2010. *http://www.eusc.europa.eu/index.php?option=com_content&task=view&id=8&Itemid=16*
[232]Richelson, Jeffery T. "Eyes on Saddam: U.S. Overhead Imagery of Iraq." Briefing book no: 88. 30 April 2003. National Security Archive 24 July 2009. *http://www.gwu.edu/~nsarchiv/NSAEBB/NSAEBB88/*
[233]Bressollette, Aurélie. "Space Current and Future Contributions to Maritime Security." Space and Internal Security Workshop. European Space Policy Institute, Vienna, Austria. 28 May 2009. *http://www.espi.or.at/images/stories/dokumente/Conference2009/bressollette.pdf*
[234]For the text of the draft Code of Conduct see *http://register.consilium.europa.eu/pdf/en/08/st17/st17175.en08.pdf.*
[235]For a detailed comparative analysis on the content of these proposals see Rathgeber, Wolfgang, and Nina-Louisa Remuss. "Space Security – A Formative Role and Principled Identity for Europe." ESPI Report 16. Vienna: European Space Policy Institute, 2009: 50 ff.
[236]Commission of the European Communities. "Space activities of the European Union – Analytical Report." Flash EB No 272 of October 2009. Brussels: Eurobarometer/Gallup Organisation: 10.
[237]*Ibid.*: 12.
[238]*Ibid.*

Exploration

1.9 Human space exploration and European identity

Gerhard Thiele

Human space exploration is frequently judged as being one of the boldest (technical) endeavours, on which humankind has embarked in its history. It is often described with very lofty terms, especially when looking for arguments, why humans engage in human space exploration in the first place. The desire and even the necessity to go beyond the known in order to be able to grow, the wish or even the imperative to expand life into the universe, are just two arguments, which are often brought forward when discussing human space exploration. From this and similar arguments one may conclude that the focus of space exploration is not only on the scientific discovery of today's unknown worlds. It is not just about enhancing our knowledge and understanding of the universe – how it looks today, how it developed and which laws govern its evolution. One could argue that besides these scientific achievements the attempt of the human to leave Earth and to travel into space is a manifestation of the "undefined striving into the open" as the German theologian Wolfhart Pannenberg describes the seemingly untiring driving force, which motivates human action.[239] Thus, human space exploration must not only be seen as an outward journey. It also needs to focus on the human being itself, as the human being appears to be at the centre of this activity. This raises the question of who the human being really is, which can be understood as a question regarding the identity of a human being.

In order to try to establish a link between human space exploration and European identity, it is first necessary to understand that the overall identity is a complex and multilayered concept. This difficulty is even reinforced in the specific case of Europe by the evolving nature of a European identity: one can not easily grasp what European identity entails at a given period. However, two elements that can be considered as constitutive of an identity are values and symbols. Human space exploration is resting on a solid corpus of values and at the same time it represents a very visible and obvious symbol. As such, European human space exploration endeavours could not only be part of the European identity, but also help further shaping it.

1.9.1 About identity

The word identity can have many different meanings depending on the context in which it is being used. Before looking into the question of the meaning of a European identity, one shall first look at the word identity itself. The Merriam-Webster dictionary defines identity (amongst other definitions) as a: "the distinguishing character or personality of an individual", hence his individuality and b: "the relation established by psychological identification." While these two definitions clearly refer to the human being as an individual, two further definitions, which seem to be more abstract, may be helpful and are therefore quoted here, namely c: "sameness of essential or generic character in different instances" and d: "sameness in all that constitutes the objective reality of a thing" or "oneness".[240]

Discussions about identity and what it could mean in a given context are manifold and the purpose of this article is not to make an in-depth review of all the various concepts. However, we interpret the definition as given above for the purpose of this article in the following way. The individual develops throughout the early years a certain personality and a distinguishing character, a process, which can be considered as essentially concluded after the puberty and the first years as an adult. This grown identity distinguishes the individual from others and coincides in the ideal case with the individual's self-awareness as a unique person. However, this grown identity must not be seen as one specific identity only. Depending to which group an individual persons feels associated with or to which group she or he is associated by others, certain other identities can be attributed to an individual. In this sense a person's identity could be seen as an accumulation of several different layers of identities, which psychology describes as a relation with others. For example, one may have an identity as man or women, husband or wife, father or mother, son, daughter, sports fan, religious (or non-religious) person and so on. Thus one may share a certain identity with others even though one has several different identities depending on the role and therefore also owns a unique identity as an individual.

How can these ideas be transposed towards the concept of a European identity? While we have a certain understanding on the concept of "identity" more questions arise when talking about a specifically European identity.

1.9.1.1 European identity

What is a European identity? This question is not so easy to answer and one may even rightfully question if a specific European identity exists. The first difficulty already arises when trying to define Europe. The German journalist Ulrich Wickert observed that despite the difficulty to describe or define what Europe is, Europeans feel very distinctly what Europe is when they are not living in Europe.[241] This points to the fact that some kind of common understanding or feeling about what Europe means, exists among Europeans, though it may not be straightforward or even possible to describe with words what this understanding encompasses exactly. Nevertheless, one can conclude that a certain European identity exists, as elusive as this concept may appear initially.

There have been various approaches trying to understand and define what the characteristic features of a European identity are, starting with the cultural heritage, the geographical landscape or the joint political entity called European Union. These various aspects have been discussed intensively in the literature, beginning with the European history, further discussing the concept of a cultural identity reflecting on language, religion or arts, and debating the various considerations from a political viewpoint like the East – West dichotomy or the Nation – European duality.[242] A prudent conclusion could be that the European identity as such does not exist, but that it may take shape in very concrete situations.

1.9.2 European symbols as part of a European identity

Almost every identity needs symbols. This becomes very obvious when looking for example at the United States. The relevance and acceptation of the national flag "Stars and Stripes" or the national anthem "The Star-Spangled Banner" becomes apparent in everyday life. Each sports event is preceded by the singing of the national anthem and the "Stars and Stripes" are ubiquitous. Thus both, flag and anthem are flown or heard even during purely national events, when no "other" party is present. Europe has similar symbols, however, they hardly play a noticeable role in the public, let alone in the life of a European citizen. The flag of Europe is far less present in the public and the European Anthem "Ode to the Joy" is heard only rarely at very selected events. This situation is mirrored in the space field as well, as illustrated by the following example. For every cosmonaut crew, which returns from a spaceflight, a traditional welcome home ceremony is held in Star City, the so-called "Vstretcha". At the beginning and at the end of the ceremony the national anthems of the returning space fliers are being played. However, if a European astronaut was part of the returning crew, the anthem of the respective nation will be played and not the European anthem.

1.9.2.1 Human space exploration as an identity symbol?

European symbols exist, but are not widely used, as mentioned above. In this respect, it is legitimate to ask to what extent human space exploration and more specifically, a visible and dedicated European engagement in and contribution to human space exploration, could become a distinct European symbol and thus contribute to the emergence of a distinct European identity. Today, if a European astronaut from a given country participates in a space mission, it is hardly noticed in the other parts of Europe. While in the astronauts' country the news coverage is usually very broad and at a prominent place in the newspaper or the TV news, in other European media the space mission usually receives hardly more than a short mention. Does this reflect a European disinterest in exploration and human activities in space? This conclusion would be premature, as the media reaction on the recent European astronaut selection showed. In its final report, ESA highlighted that the ESA astronaut selection received a European wide media coverage, which reached a depth (measured in the estimate number of people reached by the media) far beyond the coverage of most other ESA projects.[243] Interestingly, in France the announcement for the upcoming astronaut selection received more media coverage than the quasi-simultaneous flight of Leopold Eyharts, a French national, who participated in the COLUMBUS mission, which was for ESA one of the most important missions to the International Space Station (ISS) in the past years.

Marcus Hornung noted that "the knowledge about space activities is in fact very modest, although there is a generally positive public opinion towards European space efforts".[244] Humans in space are still stimulating the thoughts and feelings of the public as the coverage in all European countries on the recent ESA astronaut selection shows. One conclusion could be that the general interest for human spaceflight is still large, though it may not become visible if the goals of a mission are not known or seem to be of little interest. On the other hand, the landing of ESA's probe Huygens on the Saturn moon Titan made the front page in all major newspapers in all ESA Member States.[245] This clearly shows that an outstanding achievement of a given project, here, in the case of the Huygens probe, a scientific and technological achievement receives wide attention and high recognition throughout a broad community. For human exploration a similar success might be possible if Europe engages in a challenging technologic or scientific activity with a clearly defined goal. A cornerstone of such a European engagement could be the development of a human space transportation system. The technological challenges of such a system are well known and can be mastered by the European industry. The flight of European astronauts, who are flying into space on a European carrier, will be perceived as a significant achievement and will necessarily receive European wide attention. Initially this attention might be more pronounced in the countries of citizenship of the astronauts aboard the capsule. However, it will spread and grow if the flight of European astronauts on a European vessel does not remain a single event and instead becomes part of a larger endeavour on a regular and accountable basis. First of all, more and more countries will thus become a visible part of such an undertaking, assuming that the astronaut corps is not recruited from just a few European countries. This visibility will not develop by itself into a European identity. Visibility is a necessary but not a sufficient condition. A European human space transportation system would merely be a first step on a long road. Even more important is the contribution, which Europe is willing to provide in an international, if not global, effort in human exploration. Besides a worldwide defined and accepted goal, this requires a European commitment and persistence to participate in human exploration, which may develop even into a leading role. Leading role is here understood as engagement in an activity, where the potential leader announces credible objectives or projects, proving that he can implement them independently, initiating the central nucleation point of "accretion" around which a broader project might emerge.[246]

While the further development of European human spaceflight activities could be a strong symbol in fostering a European identity, it is also important to shed light on what is already existing, namely the European Astronaut Corps and their Charta.

1.9.3 The Charta of the European Astronaut Corps

It may be helpful examining the Charta of the European Astronaut Corps when trying to understand the role human space exploration could play in the fostering of a European identity. The Charta was signed at the European Astronaut Centre (EAC) in Cologne on 15 August 2001 by all sixteen astronauts, who composed the

European Astronaut Corps at that time. The Charta in itself may be seen as a tangible manifestation for the formation of an identity within the recently formed European Astronaut Corps.

Historically, the interest in human spaceflight activities has been different in the various European countries, leading to the formation of an astronaut corps by the European Space Agency (ESA) in 1978 and followed by the foundation of several national astronaut teams in the early 1980s, namely in France, Germany, and Italy. Besides the recruitment of national astronauts various countries have engaged in bilateral agreements with the former Soviet Union (and later on Russia), which resulted in spaceflights for an Austrian and British astronaut.[247] In 1996 it was decided to form a single European Astronaut Corps with the understanding that all national astronautical activities would eventually cease. As a consequence, this new group of European astronauts encompassed astronauts from the former national astronaut teams from France, Italy, and Germany. Together with six astronauts selected by ESA in 1992, and two new astronauts from Belgium and The Netherlands, the new European Astronaut Corps consisted of 16 astronauts from seven countries (Belgium, France, Germany, Italy, Spain, Sweden, Switzerland, and The Netherlands).

The newly formed astronaut corps, which was completed in the year 2000, had thus both a European and a "national" heritage. This resulted in noticeable differences. France emphasised from the very beginning the importance of access to space and the role of a reliable space transportation system for all other space activities. As a result the majority of French astronauts had a background as test pilots even though a human space transportation system did not yet exist in France or in Europe. In Germany's space programme more emphasis was put on science and space applications. Therefore the majority of German astronauts had a scientific background. Yet another and probably more important factor for the diversity in the new European Astronaut Corps resulted from the previous training experience and the different education as astronaut. Six astronauts had spaceflight or training experience in Star City (Russia), the others had an astronautical background through training with NASA. Initially, only three astronauts were seconded to join the NASA astronaut training after they had completed their spaceflight training in Star City. In addition, two astronauts had no training experience at all. The corps acknowledged this very diverse background and agreed early on to develop a charta, which could serve as a guideline for the united astronaut corps. The main questions to be addressed by such a document would be the following: why do we, as individuals, engage in human space exploration; how do we fulfill our duties and what are our guiding values. The Charta answers these questions and thus provides a kind of identity for the European Astronaut Corps. The vision statement mirrors the ESA values to "Shape and Share". The mission statement specifies both aspects "We shape..." and "We share..." more explicitly, thus describing generally the mission entrusted to the astronauts in their own words.

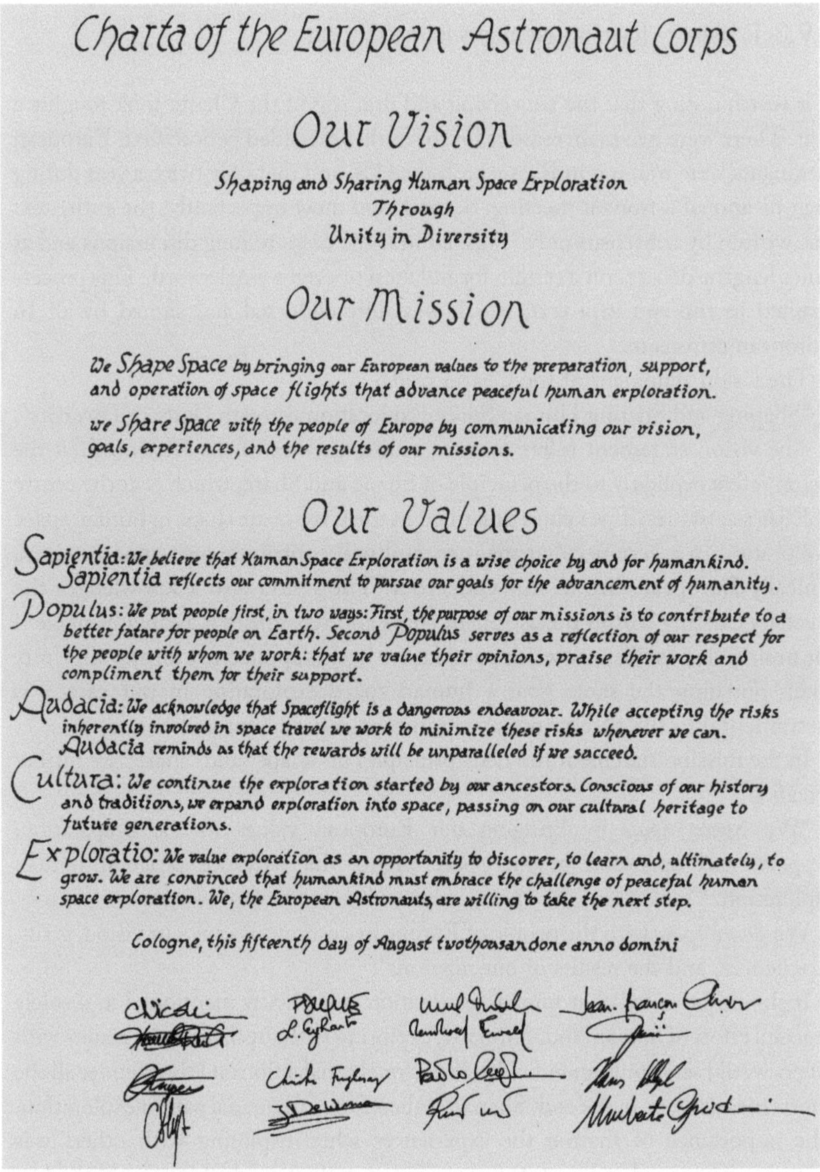

Charta of the European Astronaut Corps

Our Vision

Shaping and Sharing Human Space Exploration
Through
Unity in Diversity

Our Mission

We Shape Space by bringing our European values to the preparation, support, and operation of space flights that advance peaceful human exploration.

we Share Space with the people of Europe by communicating our vision, goals, experiences, and the results of our missions.

Our Values

Sapientia: We believe that Human Space Exploration is a wise choice by and for humankind. Sapientia reflects our commitment to pursue our goals for the advancement of humanity.

Populus: We put people first, in two ways: First, the purpose of our missions is to contribute to a better future for people on Earth. Second Populus serves as a reflection of our respect for the people with whom we work: that we value their opinions, praise their work and compliment them for their support.

Audacia: We acknowledge that Spaceflight is a dangerous endeavour. While accepting the risks inherently involved in space travel we work to minimize these risks whenever we can. Audacia reminds us that the rewards will be unparalleled if we succeed.

Cultura: We continue the exploration started by our ancestors. Conscious of our history and traditions, we expand exploration into space, passing on our cultural heritage to future generations.

Exploratio: We value exploration as an opportunity to discover, to learn and, ultimately, to grow. We are convinced that humankind must embrace the challenge of peaceful human space exploration. We, the European Astronauts, are willing to take the next step.

Cologne, this fifteenth day of August twothousandone anno domini

Fig. 1.23. *Charta of the European Astronaut Corps.*

The Charta is concluded by a set of values, which aim at reflecting typically European values. These values guide the members of the European Astronaut Corps when fulfilling their duties.

1.9.3.1 Identity forming elements in the Charta

It is worth noting that the conceiving and drafting of the Charta took roughly a year. There were two main reasons for this rather extended period: first, European astronauts were training in Russia or in the U.S. and met only twice a year during their bi-annual astronaut meeting. Second, and most importantly, the entire text was written by consensus only. This led in some cases to long discussions and at times lengthy debates on a certain formulation or even a single word. This process resulted in the end in a text, which was fully endorsed and signed by all 16 European astronauts.

The vision statement of the Charta reads:

"Shaping and Sharing Human Space Exploration through Unity in Diversity".

The vision statement refers to two guiding themes of the Charta. First the vision refers explicitly to the principle of Shape and Share, which is at the centre of ESA's activities. The values described in the Charta are shaping human space exploration in a very peculiar manner, and will contribute over time to give a typically European face to this endeavour. The vision also acknowledges the diversity existing within the astronaut corps and recognises it as an advantage if the multitude of skills, knowledge and know-how is applied to a joint goal. Unity when pursuing the same goal – human space exploration in our case – is instrumental to the success of any significant engagement and endeavour.

In the mission statement the ESA principles of "shape" and "share" are further specified:

"We *Shape Space* by bringing our European values to the preparation, support, and operation of the space flights that advance peaceful human exploration.

We *Share Space* with the people of Europe by communicating our vision, goals, experiences, and the results of our missions".

It should be noted that human exploration is explicitly mentioned as a solely peaceful effort of humankind. Similarly, exploring without sharing the results with others would serve only an individual's personal satisfaction and thus render all the efforts meaningless in the end. Sharing is therefore an essential part of exploration. The importance of sharing the experiences when exploring with others was recently emphasised in the statement of space explorers, which was published by ESPI in 2010.[248]

The process of elaborating the Charta was in itself an identity-forming element. Indeed, the set of values laid down in the Charta was agreed upon after many iterations and summarising a multitude of suggestions and proposals. It was agreed that the principle values shall be summarised and expressed in the Latin language. Combining the first letter of each value will spell the word SPACE. The values are Sapientia, Populus, Audacia, Cultura and Exploratio:

Sapientia: We believe that Human Space Exploration is a wise choice by and for humankind. Sapientia reflects our commitment to pursue our goals for the advancement of humanity.

Populus: We put people first, in two ways. First, the purpose of our mission is to contribute to a better future for people on Earth. Second, Populus serves as a reflection of our respect for the people with whom we work, that we value their opinions, praise their work and compliment them for their support.

Audacia: We acknowledge that Spaceflight is a dangerous endeavour. While accepting the risks inherently involved in space travel, we work to minimise these risks whenever we can. Audacia reminds us that the rewards will be unparalleled if we succeed.

Cultura: We continue the exploration started by our ancestors. Conscious of our history and traditions, we expand exploration into space, passing on our cultural heritage to future generations.

Exploratio: We value exploration as an opportunity to discover, to learn and, ultimately, to grow. We are convinced that humankind must embrace the challenge of peaceful human space exploration. We, the European Astronauts, are willing to take the next step.

While it is not intended to discuss all the facets of the values here, it should be mentioned that human space exploration is considered as a necessary step into our future as humankind. Consequently, the decision to participate in exploration is ultimately a wise choice as reflected in Sapientia. Furthermore, with the second value the European Astronaut Corps clearly recognises that human exploration as all other space activities must focus on the need and interest of the people not only in Europe but on the entire Earth.

This specific process of drafting a Charta for astronauts is unique in the world. It testifies of the importance for Europeans engaged in human spaceflight activities not only to be part of a universal endeavour in the name of all mankind, but also to find a specific European way to do so. Indeed, the content of the Charta reflects some of the most important values at the core of the European identity. It shows that human space exploration can help unifying Europeans around a common set of identity-forming elements, and further that Europe can share its values with the rest of the world by participating in global human space exploration endeavours. As such, the ESA principle of "shaping" and "sharing" also applies to the European identity.

1.9.4 Conclusion

The European identity as such is rather a complex process than a set of established and observable facts. As a consequence, the concept is not only difficult to grasp, but also – and more importantly – difficult to foster and encourage. Human spaceflight might be one of the few activities that have the potential to positively affect the development of a European sense of identity. Indeed, human space exploration combines both intangible values – entrenched in the Charta of the European Astronaut Corps – and tangible symbols – European astronauts floating weightlessly aboard the ISS. It is precisely the combination of these two elements that might have a strong appeal to the European citizens.

It is however obvious that the development of one aspect of a European identity as sketched above needs time, which is not measured in years but rather in decades and may very well take a full generation if not even more time. It is also unlikely that human exploration alone can achieve a strong sense of European identity if Europe is not growing together in other political fields as well. A substantial European engagement in human exploration, however, will certainly strengthen the process of building a European identity. At times it may even be the driving force behind such a process.

[239] Pannenberg, Wolfhart. Was ist der Mensch? Die Anthropologie der Gegenwart im Lichte der Theologie. Göttingen: Vandenhoeck & Ruprecht, 8. Aufl. 1995.

[340] Merriam-Webster Enzyklopeadia. 30 May 2011. *http://www.merriam-webster.com/dictionary/individuality*.

[341] Wickert, Ulrich. "Europa." In: Der Kanzler wohnt im Swimming Pool oder wie Politik gemacht wird. Schröder-Kopf, Doris (ed.). München: Droemer Knaur 2003.

[342] Strath, Bo (ed.) Europe and the Other and Europe as the Other. Frankfurt am Main: Peter Lang 2010.; Chimisso, Cristina (ed.) Exploring European Identities. The Open University 2005.

[343] ESA. ESA Astronaut Selection Promotion Campaign. ESA publication 2008.

[244] Hornung, Marcus. "European Identity through Space – How to Make Public Opinion Instrumental." ESPI Perspectives 37. 22 June 2011. *http://www.espi.or.at/images/stories/dokumente/Perspectives/ESPI_Perspectives_37.pdf*.

[245] Jean-Jacques Dordain, ESA Director General. Personal communication 2008.

[246] Bonnet, Roger. "Science Autonomy and Dependence." Presentation. European Autonomy in Space. ESPI, Vienna, Austria. 17–18 January 2011.

[247] Although the spaceflight of Helen Sharman from the United Kingdom was eventually performed as a private endeavour as the British government discontinued the project.

[248] ESPI. Human Space Exploration – A Cultural Quest around the World. October 2010. ESPI Memorandum. 22 June 2011. *http://www.espi.or.at/images/stories/dokumente/leaflet/Human_Space_Exploaration_-_A_Quest_of_Cultures_around_the_World.pdf*.

A look on critical areas

1.10 Where the drive for integration and identity fails to succeed

Wolfgang Rathgeber

Europe is a successful and important spacefaring entity. It has achieved impressive results in various areas of spaceflight, even if it does not spend huge amounts of money on space, and it has developed into a renowned international space actor. Commonly, the implementation of space activities is seen to contribute to European integration and to define, build up and sustain a European identity. However, one might also ask whether the pursuit of space activities bears an inherent risk of endangering integration and identity by offering potential for conflict and by possibly leading to or facilitating a trend towards re-nationalisation. Some pertinent issues will be discussed in the following.

1.10.1 The setting

European space activities are run within a special context mirroring the particular political set-up of Europe. There are three major levels of action. The supranational level is constituted by the European Commission, which is getting strongly involved in space through the so called flagship projects Galileo and GMES (Global Monitoring for Environment and Security) and which provides major funding, for example through its Framework Programmes for research and technological development.

The intergovernmental level is represented by the Council of the European Union and agencies like the European Defence Agency (EDA) as well as by other international organisations like EUMETSAT (European Organisation for the Exploitation of Meteorological Satellites) or the European Organisation for Astronomical Research in the Southern Hemisphere (ESO). The intergovernmental level also comprises joint efforts of single states on a bilateral or multilateral basis, often outside official European Union structures.

Last but not least, there is the national level that consists of the Member States making up Europe. In certain domains like defence, the Member States are even the principal actors. The national level is augmented by the regional level that is also getting engaged in space, for example in the NEREUS (Network of European Regions Using Space Technologies) platform. The regional level sometimes features cooperation schemes that exceed national borders.

The variety of levels is complemented by the heterogeneousness of the actors in the field. They can be grouped into public and private, military and civilian, commercial and non-profit ones with a degree of ambiguity. Endowed with different features, they also follow different goals and they are usually not coordinated. Moreover, there are various funding schemes that complicate the conduct of common projects.

An important boundary condition for European space activities is set by the political framework governing the complex interplay of all the entities involved. It is currently undergoing a process of adaptation. The European Space Policy of 2007, besides defining overarching goals and objectives, addresses the structural set-up of European space activities by formally listing the European Union as a space actor, by referring to the link between space and security and by encouraging a structured dialogue in this field. The Lisbon Treaty makes space a shared competence within the European Union and calls upon the Union to draw up a European space policy.

1.10.2 The case of ESA

ESA is one of the better known examples of European space actors. Its set-up is quite unique and deserves some attention. Being an international organisation, it is composed of 18 Member States, featuring a constituency different from the one of the European Union. ESA runs programmes that are mandatory for each Member State – the science programme and the general budget execution. These activities are funded by national contributions that are based on each country's gross national product. In addition, ESA runs a number of optional programmes where Member States can decide whether and how much they contribute.

A specific feature of ESA is the principle of fair return (or "juste retour"). This principle, while still incorporating competitive elements, entails the procedure of granting industrial contracts to each Member State in relation to the financial contribution of the specific Member State. Thus, it offers an incentive to the Member States to provide funds. The principle of fair return makes ESA funding different from EU funding and is considered one of the reasons for ESA's continuing success.

Fig. 1.24. *European Spaceport (credit: ESA).*

This success has been evident in many ways and continues to show up through a variety of ongoing and planned projects. The science and exploration missions of ESA constitute outstanding scientific and technological endeavours. Later examples include the Herschel and Planck missions. The Herschel spacecraft carries the largest and most capable infrared telescope ever used in space, studying the origin and evolution of stars and galaxies.[249] Planck employs detectors that are kept at temperatures close to absolute zero to observe fluctuations of the cosmic microwave background.[250] These undertakings have augmented and sustained Europe's prestige as a significant and capable space actor.

ESA is also responsible for the Columbus module and the Automated Transfer Vehicle (ATV), the European contributions to the International Space Station (ISS). The ISS is the biggest space structure ever created by mankind, and being part of this global project gives Europe a high degree of visibility on the international scene. Moreover, Europe has its own access to space through its economically successful and technically reliable Ariane launcher, soon to be complemented by the Soyuz and Vega vehicles.

Beyond these tangible achievements, there are political meta-goals that have been reached through ESA. International cooperation within Europe has been facilitated and supported by providing common and attractive targets. This cooperation has led to a consolidation of a pan-European spirit by creating a sense of shared pride. International cooperation triggered by ESA, however, has not only taken place within Europe. In fact, ESA is the only space actor that cooperates with all other space actors worldwide. In addition, ESA also cooperates with non spacefaring nations and developing space actors. This allows for the exploitation of global synergies and it supports overarching goals of European foreign policy.[251]

Last but not least, the industrial policy of ESA has contributed to the emergence and stable existence of a sound and competitive industrial base for space in Europe, comprising global players as well as highly specialised entities in designing, manufacturing and systems integration. The European industry has proven to be able to implement space programmes at low cost, even though it suffers from a relatively low funding level compared to the United States that features a significantly higher amount of institutional spending.[252]

1.10.3 Stories of slow progress

Alongside these success stories, Europe is also facing challenges in the domain of space. The two flagship projects of Galileo and GMES are a marked example. Galileo, as a European space-based navigation system, is meant to free Europe from dependence on the U.S. for space-based positioning, navigation and timing services. While the technical aspects of Galileo are relatively straightforward, the political and economical dimensions have proven to be hard to tackle. Even before the actual implementation, negotiations with the U.S. and other international partners were complex and lengthy.

The overall governance scheme of Galileo still remains to be settled in detail, even if some operational aspects have been clarified by now. This is complemented by a difficult purchase scheme. A first procurement process failed due to a complex situation on the bidders' side, and the two main stakeholders ESA and the European Union had to re-organise the division of labour and responsibility. This happened against a massive cost overrun with the European Parliament having set a budget ceiling of 3,4 billion euros.[253]

All of this has led to an undue delay. Initially foreseen to enter into service by 2010, Galileo is now likely to be fully operational only by 2016.[254] This, in turn, potentially endangers frequency utilisation agreements concluded with the U.S., because China might complete its Beidou system before Galileo, securing itself certain portions of the electromagnetic spectrum in the process. Moreover, time has been lost in defining Galileo's link to the security and defence domain. Galileo had been conceived as a civilian system, but its inherent dual use potential has raised the question of how to involve military entities that

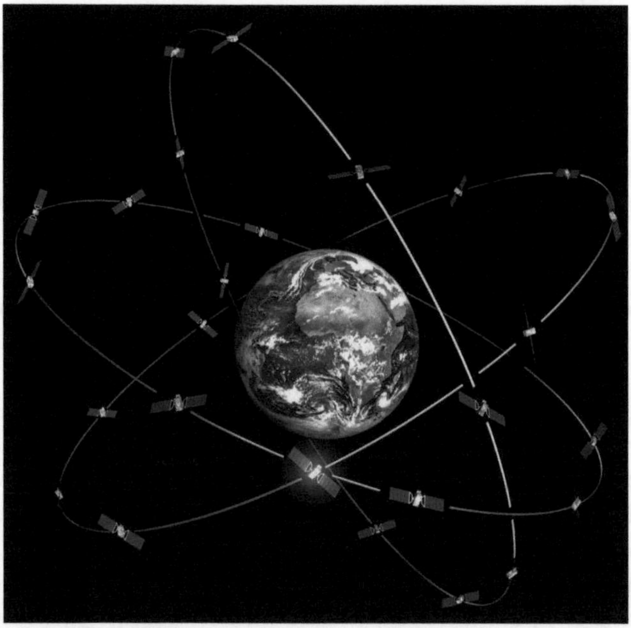

Fig.1.25. *Galileo Constellation (credit: ESA).*

could profit from a genuinely European navigation capability. Putting everything together, the difficulties with Galileo mirror the difficulties of the European set-up for space activities where many actors are involved, roles are not clearly assigned and where national or industrial interests endanger the execution of common projects.

GMES is the other flagship project. It is meant to bring together both space-based and in-situ measurements of Earth observation with a view to offer suitable products and added-value services. Again, the concept is technically clear, but hard to grasp. Relevant markets need to be developed. Whereas Galileo can be seen and sold as the European answer to GPS (Global Positioning System) and can be symbolised by handy and familiar reception devices, GMES is considerably harder to bring to people's minds. Accordingly, it has been called an animal without a face. It seems that the problems around GMES are mostly related to communication.[255] This has been aggravated by latent naming issues that touched upon legal questions and matters of national prestige.

Again, GMES features a link to the security domain. After all, it contains the letter S for security. There are also considerations for a dedicated security satellite within GMES. The security exploitability of GMES needs to be defined and delimited as part of a wider European approach to the issue area of space and security that will be discussed below. For example, contributing missions that are often run at national level will have to be integrated, which causes the need for interoperability and a tailored data policy. Finally, GMES also faces financial difficulties, and ESA has made it clear that it is not willing to dedicate more funds than already committed.

As mentioned, the area of space and security is another challenge for Europe. While Member States have been active in this regard for some time and have achieved impressive results by running their own systems in Earth observation, communication and signal intelligence, Europe as a whole has been reluctant to embark on this domain. The European Space Policy of 2007 contains a separate chapter on security and defence and encourages a structured dialogue of relevant entities. European institutions like EUSC (European Union Satellite Centre) or EDA are important players in the domain of space and security. Still, as laid down, the bulk of security-related space activities is carried out at national level. The corresponding systems are optimised for national needs, not for European ones. Usually, they are run by military entities that are reluctant to pass on data and information. There are cooperation schemes based on complementarity at bilateral or multilateral level, but these schemes are often set outside official European structures.

A true Europeanisation of security-related space activities will require the political will of Member States to pass on part of their national competencies to a European level. This will take a while, because security and defence matters are at the heart of national sovereignty. Once the political will has become manifest, the technical implementation of interoperability standards and an exchange

Fig. 1.26. *European Union Satellite Centre (credit: EUSC).*

scheme codified in a suitable data policy will have to follow. Besides, the definition, set-up and operation of security-related space systems will have to be run jointly. First steps have been taken for Earth observation in the framework of MUSIS (Multinational Satellite-based Imagery System for Surveillance, Reconnaissance and Observation), but progress has been slow due to national interests that seem to be perceived as contradictory to European ones.[256]

The issue area of space and security is an important test case for the European integration. Besides, it can constitute a facilitator for a stronger political sense of European identity. Commonly pursued goals will raise awareness of a truly European actorship and responsibility, illustrating also that European cooperation can unleash synergy potentials that single Member States do not dispose of on their own.

There are numerous upcoming projects that can serve as incubators. For example, in the area of Space Situational Awareness (SSA), Europe strives for a higher degree of autonomy as well. SSA comprises the knowledge about the orbit population, the location, motion, function and status of spacecraft as well as about the threats they face. These data largely had to be retrieved from the U.S. so far. At the Ministerial Council of 2008, ESA decided to set up its own SSA system linking existing national and European facilities with new ones to be built. The European SSA initiative will have to involve military entities, accounting for their interests and having them contribute to the communitarian endeavour by shouldering common burdens and by supplying necessary information. Other examples include the European Data Relay Satellite (EDRS) supporting fast transmission of Earth observation data or the Global Integrated Architecture for Innovative Utilisation of Space for Security (GIANUS).

1.10.4 What's wrong?

Given the situation sketched above one might ask what the reasons for the present state of affairs are. In general, there is a lack of awareness of the importance of space and not enough knowledge about the potential space holds for various specific and overarching policy objectives. This lack of awareness or concernedness, respectively, translates into a lack of political momentum that would be needed to push space forward. Moreover space, unlike other common projects, is not sufficiently seen as a European endeavour. This might be linked to the emergence of space activities in the Cold War era, where they expressed national prestige and capability of the two superpowers of the time and where Europe was not part of the game.

In addition, in spite of ongoing adaptations, the European set-up of institutions is not yet optimised with regard to the efficient conduct of space activities.[257] The multi-level approach can be seen as an asset, but it also presents the risk of needless duplication or even negative interference. Entities like the EUSC dispose of high expertise and of operational routine, but their full capacity is only getting started to be used. The division of work and responsibility between the European Commission and ESA has been addressed, for example in the corresponding Framework Agreement, but it will have to be refined further and to be constantly adapted.

Also within the specific levels, the presence of heterogeneous actors sometimes presents a non-interfering co-existence at best. Within the Member States, military and civilian entities do not always communicate properly. Whereas the civilian institutions, especially in the public domain, strive for transparency, the militaries traditionally stick to a "need to know" culture that hinders a free flow of information and that prevents the unlocking of synergies. While certainly justifiable in the case of national security interests, there might be excessive degrees of confidentiality in other, less critical issue areas.

As for the interaction between the single Member States, a growing rivalry and a reinforced trend towards re-nationalisation can be observed. Space activities as such do not drive this development. Instead they just make it evident. But their visibility and their symbolic content could contribute to consolidating the trend. Therefore, it is important to free space activities from the aura of re-nationalisation. This leads directly to measures that need to be taken to enhance the role that space can play for shaping a European identity and for driving further European integration.

1.10.5 The way forward

First of all, the link between European identity and space needs to be established. This of course raises the question of what a European identity encompasses. A European identity arising from Europe's sheer nature is a matter of intense debate. From a functional and political perspective, however, the European identity shows up in basic principles and values as they are encoded in the Treaty of Maastricht or the European Security Strategy. These principles and values comprise the promotion of peace, democracy, human rights and the rule of law. Typically, Europe tries to pursue these goals by using means of diplomacy, multilateralism and a combination of civilian and military instruments, the latter constituting the last resort only. This makes the European approach quite unique. Refraining from

Fig. 1.27. *UN General Assembly (credit: United Nations).*

the traditional hard power behaviour, it can employ new dimensions of political and access strategic avenues that are closed to other actors.[258]

Space as such does not constitute a unique element of European identity, because it is global by nature. However, the way Europe utilises space can underline European identity. Using the negotiation potential of space, cooperating with many different space actors and carefully being aware of the dual-use nature of space, Europe applies its typical tools in space and through space. In line with its soft power approach, Europe also shows political leadership in efforts to oppose the weaponisation of space. The Draft Code of Conduct for outer space activities[259] presented in late 2008 and due to be taken further by the international community is a marked example. In this regard, Europe has not just reacted to a political agenda set by others, but has exerted its influence in a proactive way.

Also in a more general way, the handling of space by Europe just mirrors its internal structure in a prominent way. The issue area of space and security is probably the most illustrative showcase. The underlying question is how many national competencies Member States are willing to transfer to the European level, and to what degree European tasks will be centralised. If Europe agrees to communitise security and defence matters, it will have to run the corresponding programmes in a truly coordinated and coherent way that leads to an utilisation of complementary abilities. A suitable implementation of space activities reflecting the overall scheme will then follow naturally.

Moves towards a higher degree of European integration should be flanked by augmented outreach activities and enhanced education efforts. Space should regain its status as a subject of fascination and of pride, especially among young people. In addition, it should be made clear that space, beyond its inherent features, is a valuable tool to reach overarching political goals. All of that, besides political will, needs sufficient and sustainable funding. This involves enough money provided by the Member States as well as justifiable and efficient allocation mechanisms. In this regard, ESA's fair return principle can be used as an inspiration and as an element of an attractive, successful and viable set-up. Europe should try hard and reach out for space – after all, there are stars on its flag.

[249] "Herschel Science Objectives". 2009. European Space Agency 2 February 2010. *http://www.esa.int/SPECIALS/Herschel/SEMSN00YUFF_0.html.*

[250] "Planck Science Objectives". 2009. European Space Agency 2 February 2010. *http://www.esa.int/SPECIALS/Planck/SEM0P20YUFF_0.html.*

[251] Rathgeber, Wolfgang and Nina-Louisa Remuss. "Space Security. A Formative Role and Principled Identity for Europe." January 2009. ESPI Report 16. 20 January 2010. *http://www.espi.or.at/images/stories/dokumente/studies/espi%20report%2016.pdf.*

[252] Rathgeber, Wolfgang. "Space Policies Issues and Trends in 2008/2009." May 2009. ESPI Report 18. 19 January 2010. *http://www.espi.or.at/images/stories/dokumente/studies/espi%20report%2018.pdf.*

[253] de Selding, Peter. "European Commission To Trim Initial Order of Galileo Satellites." October 2009. Space News. 20 January 2010. *http://www.spacenews.com/policy/european-commission-trim-initial-order-galileo-satellites.html.*

[254] *Ibid.*

[255] Rohner, Nicola, Schrogl, Kai-Uwe and Simonetta Cheli. "Making GMES Better Known. Challenges and Opportunities." Space Policy 23 (2007): 195–198.

[256] "MUSIS Ground System Deal Teeters on Edge of Collapse". Space News 26 April 2010: 6.

[257] Rathgeber, Wolfgang. "The European Architecture for Space and Security." August 2008. ESPI Report 13. 21 January 2010. *http://www.espi.or.at/images/stories/dokumente/studies/espi_report_13.pdf.*

[258] Rathgeber, Wolfgang and Nina-Louisa Remuss. "Space Security. A Formative Role and Principled Identity for Europe" January 2009. ESPI Report 16. 20 January 2010. *http://www.espi.or.at/images/stories/dokumente/studies/espi%20report%2016.pdf.*

[259] Council of the European Union. "Draft Code of Conduct for outer space activities." Doc. 16560/08 of 3 December 2008. Brussels: European Union.

CHAPTER 2

PERSPECTIVES FROM THE HUMANITIES

Arts and social sciences

2.1 References on space and European identity in the arts and the social sciences

Giulia Pastorella

2.1.1 Introduction

The link between European identity and outer space might not be as evident as the link between, say, European identity and history, European identity and culture, European identity and religion and so on. Yet, throughout history there are many diverse references to such a connection, ranging from philosophers to politicians, from novel writers to scientists.

Historically, the sky has been first the realm of Gods, then the ultimate ambition of scientists, nowadays a political and economic chessboard, and it will always be the subject of human dreams. It should be of no surprise then that an endeavour to which religion, science, politics and emotions are so tightly connected should influence deeply any feeling of identity. This is so both at a national level and, as the articles published in this book have clearly shown, on a European level. Studying and, more recently, exploring space increase people's sense of cohesion, for it engenders a sense of common pride, the feeling of being on a mission, a unifying strive towards knowledge and discovery.

The aim of this contribution is to illustrate the wide variety of references to this link through a series of quotations taken from the arts and from the social sciences. Each quotation will be briefly explained by giving the context and some biographical details, as the changing conception of outer space are often the expression of a historically specific conception of space and especially of space exploration. The common thread is the European strive to explore and understand the rest of the universe which contributes to unite people.

2.1.2 Arts

The close relationship between Art and space is often being highlighted. In a recent essay for example, Ulrike Landfester analyses this trans-disciplinary discourse and applies it to three central questions: can Art be used to communicate with alien life forms, what is the impact of the physical differences between the Earth and extraterrestrial environments on human understanding and practicing of Art and finally is it possible for Art and space technologies to enter into a closer, mutually beneficial relationship.[1] The first section will be dedicated to quotations taken from the Arts. The category Arts should not be understood narrowly. In this context it means mainly the humanities, but also includes performing arts, humoristic writings, sketches etc. The concepts of space and Europe are relatively new compared to the whole history. Synecdoches and metonymies will be therefore often used in the quotations. Broader, more ancient ideas, such as the sky, the stars will stand for space, while precursors, such as Greece, will represent modern Europe. Quotations are listed in a chronological order.

2.1.2.1 A Greek on the Moon

"*The ones who gave me the biggest laugh of all were those who were disputing about boundary-lines [. . .] In fact, since from up there the whole of Greece then looked to me four fingers in size, I reckon proportionately Attica was a very tiny fraction. So I thought how little was available for these men to pride themselves on: for it seemed to me that the biggest landowner among them had just about one of Epicurus' atoms to cultivate*".

Lucian of Samosata, "Icaro-menippus, an aerial expedition", around 160 A.D.

Menippus, a Greek man interested in astronomy, is unsatisfied by the contemporary scientific knowledge on the matter. He thus decides to enquire further. Using a wing from an eagle and a wing from a vulture, he sets off for an exploratory journey to the Moon and to the Heavens. From up there, he comments on the tiny dimension of Greece and the trivial nature of man's disputes, conflicts and dramas. Space exploration allows putting things into perspective, and to overcome divisions amongst people. Anyone who has been in space should therefore exhort for a more universal, peaceful approach. Wars amongst the polis in the cradle of Europe were as senseless as wars amongst modern European countries were in the following centuries. Space helps understanding that we are all part of the same, small but important piece of the world.

2.1.2.2 The place where men's wits are kept

"Tis true to journey further ye will need,
And wholly must you leave this nether sphere;
To the moon's circle you I have to lead,
Of all the planets to our world most near [. . .]"

[. . .] A place wherein is wonderfully stored
Whatever on our earth below we lose.
Collected there are all things whatsoe'er,
Lost through time, chance, or our own folly, here.

Ludovico Ariosto, "Orlando Furioso", 1532, LXVII and LXXIII Gustav Doré

In the Italian epic poem, Astolpho flies on a hippogryph to the Moon to find insane Orlando's "sense", meaning his wisdom, judgement. If men's wisdom is on the Moon, it means that on the Earth only foolishness is left. Indeed Ariosto set the poem in 8th century, a period which saw Europe foolishly ravaged by the war between Charlemagne and his Christian paladins, and the Saracen army. Transposed in a modern perspective, this quest for wisdom on the Moon could be read as a metaphor. By going into space Europeans, and more generally the world, will understand that peace is the only desirable and sensible international situation.

2.1.2.3 A view from Sirio

«The sovereign states of Germany or Italy, which one can traverse in a half hour, compared to the empires of Turkey, Moscow, or China, are only feeble reflections of the prodigious differences that nature has placed in all beings.»

Voltaire, « Micromegas », 1752

Micromegas lives on Sirio. From his star he sets off on a journey first to Saturn, and then to the Earth. In his philosophical tales Voltaire often uses "strangers" or "naifs" characters to take a different point of view on everyday matters (e.g., "L'ingénu" or "Candide ou l'optimisme"). Using an alien allows "to have a quick overview of all centuries and countries, and therefore to see at once all of humans' silliness".[2] In this case Micromegas' huge size makes the differences amongst European countries look irrelevant, once again. Those countries, however, differ from Turkey, Russia and China, not only in terms of size but of nature, as the rest of the philosophical tale shows. While the meaning of this quotation could be interpreted in a similar way to Lucian's quotation, we should remember that at Voltaire's time there was already the feeling that Europe was somehow different from the rest of the world, unique and should for this reason be united.

2.1.2.4 Europe's confused, discombobulated, disorganised reaction to a space adventure

"Russia paid in as her contingent the enormous sum of 368.733 roubles. No one need be surprised at this, who bears in mind the scientific taste of the Russians, and the impetus which they have given to astronomical studies – thanks to their numerous observatories.
France began by deriding the pretensions of the Americans. The moon served as a pretext for a thousand stale puns and a score of ballads, in which bad taste contested the palm with ignorance. But as formerly the French paid before singing, so now they paid after having had their laugh, and they subscribed for a sum of 1.253.930 francs. At that price they had a right to enjoy themselves a little.
Austria showed herself generous in the midst of her financial crisis. Her public contributions amounted to the sum of 216.000 florins – a perfect godsend.
Fifty-two thousand six-dollars were the remittance of Sweden and Norway; the amount is large for the country, but it would undoubtedly have been considerably increased had the subscription been opened in Christiana simultaneously with that at Stockholm. For some reason or other the Norwegians do not like to send their money to Sweden.
Prussia, by a remittance of 250.000 thalers, testified her high approval of the enterprise.
Turkey behaved generously; but she had a personal interest in the matter. The moon, in fact, regulates the cycle of her

(*continued*)

years and her fast of Ramadan. She could not do less than give 1.372.640 piastres; and she gave them with an eagerness which denoted, however, some pressure on the part of the government.
Belgium distinguished herself among the second-rate states by a grant of 513.000 francs – about two centimes per head of her population.
Holland and her colonies interested themselves to the extent of 110.000 florins, only demanding an allowance of five per cent discount for paying ready money.
Denmark, a little contracted in territory, gave nevertheless 9000 ducats, proving her love for scientific experiments.
The Germanic Confederation pledged itself to 34.285 florins. It was impossible to ask for more; besides, they would not have given it.
Though very much crippled, Italy found 200.000 lire in the pockets of her people. If she had had Venetia she would have done better; but she had not.
The States of the Church thought that they could not send less than 7040 Roman crowns; and Portugal carried her devotion to science as far as 30.000 cruzados. It was the widow's mite – eighty-six piastres; but self-constituted empires are always rather short of money.
Two hundred and fifty-seven francs, this was the modest contribution of Switzerland to the American work. One must freely admit that she did not see the practical side of the matter. It did not seem to her that the mere despatch of a shot to the moon could possibly establish any relation of affairs with her; and it did not seem prudent to her to embark her capital in so hazardous an enterprise. After all, perhaps she was right.
As to Spain, she could not scrape together more than 110 reals. She gave as an excuse that she had her railways to finish. The truth is, that science is not favorably regarded in that country, it is still in a backward state; and moreover, certain Spaniards, not by any means the least educated, did not form a correct estimate of the bulk of the projectile compared with that of the moon. They feared that it would disturb the established order of things. In that case it were better to keep aloof; which they did to the tune of some reals.
There remained but England; and we know the contemptuous antipathy with which she received Barbicane's proposition. The English have but one soul for the whole twenty-six millions of inhabitants which Great Britain contains. They hinted that the enterprise of the Gun Club was contrary to the "principle of non-intervention." And they did not subscribe a single farthing".

Jules Verne, "From the Earth to the Moon", 1865

Written when space exploration was still a dream, this 19th century novel has often been defined as the first real science-fiction literary work. From the purely scientific point of view it is amazingly accurate in its conjectures, but also its political analysis approaches – a not-very far back reality. The book recalls the American project of sending a man in a cannonball to the Moon. This proposal provokes very diverse reaction in Europe. In an uncoordinated response, each state behaves following its "temperament", its national identity. It is interesting to notice how 200 years ago the national clichés were similar to those that we still have now. Of course this is an exaggerated scenario, but it is also a possible extreme result of a lack of a common European attitude. At Verne's times, Europe was not mature for a common space mission, but now it should be united enough to avoid such an embarrassing scene.

2.1.2.5 Some myths are universal!

"No one can compare the Chinese legend with the popular European belief in the 'Man in the Moon,' without feeling convinced of the certainty that the Chinese superstition and the English nursery tale are both derived from kindred parentage, and are linked in this relationship by numerous subsidiary ties. In all the range of Chinese mythology there is, perhaps, no stronger instance of identity with the traditions that have taken root in Europe than in the case of the legends relating to the moon".

W. F. Mayers. Notes and Queries on China and Japan. Hong Kong, 1869, p. 123

W.F. Mayers was secretary of the English Legation at Beijing. He studied the Chinese civilisation and culture in depth, and found out common legend, myths and modes of faith and worship. The one relating to the Moon is just an example of space-related myths (on the sky, the stars, constellations etc.[3]) which are surprisingly similar in Western and Eastern cultures. Mayers thinks this way could somehow establish the consanguinity of the nations. The idea of space as a source of universalistic feelings is not new but always powerful. Every civilisation has had the idea that the sky played an important role in our life, and it still does today, though in a different way. Europe could play on that to further its role on the international space scene, playing on what is common and not what is different.

2.1.2.6 Space-bound song

"*We're leaving together*
but still it is a farewell,
and maybe we'll come back to Earth,
who can tell"

The final countdown", EUROPE, 1986[4]

On a lighter, nearly folkloristic note, let us refresh our pop-culture memory. The very famous song "The final countdown", soundtrack to many sport events and other public occasions, was sung by a band called EUROPE. The lyrics, inspired by David Bowie's song "Space Oddity", speak for themselves. The fear and excitement of any space adventure can be felt through the powerful music. "Leaving together", going into space as a United Europe, should make the adventure even more fascinating.

2.1.2.7 Europe or ketchup?

"*Although from the outer space Europe appears to be shaped like a large ketchup stain, it actually consists of many small separate nations, each with a proud and ancient tradition of hating all the other ones.*"[5]

Dave Barry, "Dave Barry's Only Travel Guide you'll ever need", 1991

Along the same lines of "Icaro-menippus" and "Micromegas", this American humour columnist underlines the ridicule size of Europe seen from outer space. Behind this apparently light comment lies once more the serious idea that space activities might contribute to annulling the deeply rooted hatred among European countries. Peace is at the very heart of the European construction. Europeans might realise, by physically or ideologically seeing things from the outer space, that they are little more than a "large ketchup stain", so there is no need to fight amongst each other.

2.1.2.8 Well invested money . . . ?

"Europe has increased its space budget by 20%"
"Given the mess that there is down here, it is certainly about the right time we hasten space conquer"

"Yeah, let's go away from here . . . "[6]

As Michael Moore had pointed out in one of his conferences,[7] when things go badly, Americans in particular, but governments in general, turn to space exploration to channel people's attention elsewhere. Of course Europe has not invested more in space because of that, but certainly having a united, coherent and well functioning European space policy will contribute to diminish the feeling that Europe is still very divided. Also, a common space policy is much more impressive for an average European citizen than a common agricultural policy (which is so often criticised and misunderstood). The more money invested in European space policy, the more the feeling of being a real union of countries going "out there" should grow.

2.1.2.9 New Eurovision participants!

> *"Hello Earth. Greetings from the ISS. We are the only extraterrestrial team represented in the contest. Today we space travellers are going to give orders from the orbit to the Earth rather than taking them. "Europe: start voting now!!"*[8]
>
> Eurovision 2009 – Final – Signal To A Vote From International Space Station

Eurovision, a very well known song contests, has been defined as "strongly embedded into Europe's collective mind".[9] It has been broadcasted for the first time in 1956. It was an innovative experiment in live television. It has always had great success thanks to its adaptability to new technologies and increasingly international audiences. In 2009 it was decided that the official start to voting for the best song was to be given by the crew of the International Space Station. Bringing astronauts into the European contest is a symbol: space is now part of our everyday reality.

2.1.2.10 Communicating space successes

Malta

Sweden

Netherlands

Latvia

Belgium

Norway

Andorra, Spain Lithuania Bulgaria

France Italy Denmark

European stamps with astronomy and space themes[10]

This brief selection of stamps illustrates the will to raise European people's awareness of present and past space achievements. They were mainly issued in 2009, for the International Year of Astronomy, but other examples from previous years can also be found. The small size of stamps does not reflect the great ideological meaning they have. A stamp is something of everyday use, but something that people often collect and keep for decades. Stamps have a symbolic value as they help communication across national borders, they contribute to shorten distances. All important events in recent history have been celebrated by issuing a series of stamps. It seems logical that the entry of Europe into the Olympus of space powers should also be celebrated accordingly.

2.1.3 Social sciences

The second section, presents quotations mainly by politicians, but also sociologists, philosophers and political institutions. The general trend that can be evinced from the list is that space policy does indeed contribute greatly to the construction of a European identity. This section comprises more direct references to space policy than the previous one, but also indirect hints from a variety of actors. As with the previous section, quotations are listed in a chronological order.

2.1.3.1 The Council of Europe's "Celestial" choice

> *«Against the blue sky of the Western world, the stars symbolise the peoples of Europe in a circle, the sing of union. The number of stars is invariably twelve, the figure of twelve being the symbol of perfection and entirety.»*[11]
>
> Council of Europe. Paris, 7/12/1955

Although not specifically related to space policy as such, the choice of the European flag and its starry circle is not accidental. A flag is by definition a symbol that promotes the European image and identity to citizens and the world. The current one, designed by Arsène Heitz and Paul Lévy in 1955, was first adopted by the Council of Europe, and only in 1985 by the then European Community. The strong appeal of the circle of gold stars won over the other options, as it represented everlasting solidarity and harmony between the peoples of Europe.[12] Just like the twelve signs of the zodiac represent the whole universe, the twelve gold stars stand for all peoples of Europe. To 21st century people the flag also can easily suggest a reference to space exploration. Space could unite European peoples even more.

2.1.3.2 Ethics and utility conjoined in space policy

> *The launching of a Euroluna (Euromoon) from a European space organisation would have, in his words, "a first order importance, both moral and practical, for all countries of the continent"*[13]
>
> Letter from Amaldi to L. Crocco, 16 December 1958,
> Amaldi archive, Rome, box 212, folder 6.

Edoardo Amaldi was one of the founding fathers of European space policy. He realised that there was an urgent need to establish an official and independent European space organisation on the model of CERN (of which he was amongst the creators), otherwise "in twenty years we shall find ourselves separated by an insuperable gulf between those countries that can launch space vessels and those that cannot".[14] He also understood that space policy could "encourage" or even "increase" European unity "from without".[15] This led to the creation of ESRO, the precursor of ESA, setting the foundations for what is now the European Space Policy. Right from the beginning, promoters of a European space policy were aware of the moral impact it could have on European identity. In fact his approach to space policy has often been described as a philosophical one, an idealist which thought that any international initiative of men of goodwill was worth trying.

2.1.3.3 Past and present discoveries

«A cette époque-là, dit-il, l'horizon était la découverte du Nouveau Monde; aujourd'hui, c'est la conquête et l'exploration de l'espace.»
"In those days, the horizon was discovering the New World; nowadays, it is conquering and exploring space"[16]

Louis Armand, quoted in the article "Glass-Steagall Act" by A-J Holbecq

From 1958 to 1959, Louis Armand co-managed the European Atomic Energy Commission (Euratom). The parallel he draws between exploration in the classical sense, and exploration in the outer space sense shows once more that, even though the geographical areas explored might change, the essence remains that Europe is naturally prone to discoveries. Their terrestrial or celestial nature is a mode of the substantial quest for understanding the world around us.

2.1.3.4 First signs of awareness of the need for a United European Space Policy

«Il apparaît difficile de formuler des programmes futurs concernant les activités spatiales tant que l'Europe n'aura pas établi dans ce domaine, une politique unifiée et cohérente»
«It seems difficult to frame future space activities programmes until Europe has not established a united and coherent policy in that domain».[17]

(*continued*)

Report of ELDO (European Launcher Development Organization), 1960–1965, p. 6

As early as 1964 there was an attempt to build a European space project that would be a symbol for Europe and for the rest of the world. ELDO, which merged with ESRO to become ESA, was the result. It was set up to build the space launch vehicle called Europa. Already at that time, decision-makers were aware that without a strong political will behind the European project, no coherent and cohesive space policy could ever be made. A one-voice approach was, and still is, deeply needed.

2.1.3.5 Jacques Delors was right, as usual

«L'Europe ne fournit pas assez cette image de politique réelle dont se prévaut une grande puissance. Elle devra être plus universelle, ce qui lui coûtera plus cher» (Jacques Delors).[18] *C'est pourquoi, notamment, elle ne doit pas renoncer au chemin de l'ambition spatiale. Sans sacrifier aux facilités des programmes de prestige, hors de propos dans la conjoncture économique mondiale, il est de sage politique qu'elle joue de ce mythe mobilisateur, l'explorateur spatial artisan et conquérant de nouveaux mondes.»*

"Europe does not provide enough that image of real politics which a great power should use. It will have to be more universal-and that will be the costly" (Jacques Delors). This is why, in particular, it should not give up its space ambition. Without giving in to attractive prestigious projects ignoring the world economic situation, it should play on the power of the myth that will stir people's imagination: a space explorer who is at once a craftsman and a conqueror of new worlds".[19]

Jacques Delors, probably the most famous European Commission President, wanted a united, great Europe, one that would be taken into consideration in any "universal" project. Of course he was mainly referring to international politics, but this could well apply to space. Playing on the attractiveness of space activities, especially human space flights, is one of the possible ways for Europe to catch people's attention. Space as "myth" reminds us of the nearly transcendental character of some space endeavours, and of their universal appeal, like myths. However, a common space dream should not imply loosing grasp with reality. Space has a double nature: a potential market and an ideal realm.

2.1.3.6 Europe gets the chance to see itself

> *„das entscheidende Ereignis der Raumflüge besteht nicht in der Erreichung der fernen Regionen des Weltalls oder des fernen Mondgeländes, sondern darin, daß die Erde zum ersten Mal die Chance hat, sich selbst so zu sehen, sich selbst so zu begegnen, wie sich bisher nur der im Spiegel sich reflektierende Mensch hatte begegnen können"*.
> "The most important aspect of space flights consisted not so much in reaching out to the distant regions of the universe or the lunar area, but in that the Earth has for the first time the chance to be seen and see itself as a whole. Up to now only people had been able to "meet" with themselves in such a way, by seeing their reflection in the mirror".[20]
>
> Günther Anders, *Blick vom Mond: Reflexionen über Weltraumflüge,* 2nd ed. München, 1994, p. 12

Günther Anders was a German philosopher and journalist who developed a philosophical anthropology for the age of technology. He thought technology is our fate, but he was at the same time very critical towards it. Like other quotations before, this one underlines once more that space exploration brings a radical change in self-perception. While Anders was referring mainly to the Apollo mission and the U.S., space exploration had an influence on any country of the world. A change in the European "Weltanschauung" has taken place when Europe started to see itself from "out there". We have the confirmation of what already Lucian of Samosata, Voltaire and the other writers had imagined.

2.1.3.7 In search of Europe's autonomy in space

> *Galileo will mean that the EU "would not have to accept Europe's subjugation in space matters".*[21]
>
> Jacques Chirac, 26/03/2002

France's ex-President Jacques Chirac sees space as an opportunity for Europe to reinforce its international status and acquire more independence. Often the EU has been blamed for being dependent on and in the wake of the U.S. in space matters. Galileo, the endeavour that has caught the most attention by the public- unfortunately mainly for its ups and downs- is meant to put an end to this. The new satellite system, once operative, will presumably strengthen the feeling that Europe can stand on its own feet, as a united political entity, in the space field but also on the international scene. Despite Europeans' reluctance to become a so called "hard power", certainly some more independence will be welcomed as a sign of maturity of the Union.

2.1.3.8 Towards a hybrid Euro-national space policy?

> "*Like all the other space powers, France and Europe must use the space sector to confirm both their expertise and respective identities*".[22]
>
> Christian Cabal, Member of the French Parliament, and Henri Revol, French Senator

The interaction between national and European identity, which is reflected in the interaction between national and European space institutions and programmes, is still the focus of a heated debate. Compatible, incompatible, substitutable, additional are only some of the options for the interaction foreseen. The French Parliament is here underlining the need for coordination between the two levels, especially given the importance of the French space sector on the European scene. The complementary nature of the two levels is implicitly stressed in the sentence, but so is the will to keep the French identity – and France's space policy – as a separate entity from the European one.

2.1.3.9 The European Commission's point of view

> "*Space research [. . . .] is essential to Europe's identity and leadership as a knowledge based society*"
>
> WHITE PAPER Space: a new European frontier for an expanding Union "An action plan for implementing the European Space policy" Brussels, 11 November 2003.
>
> «*Space can contribute to European cohesion and identity, reaching citizens across all countries*» and «*The international exploration endeavour has a significant*

(*continued*)

political appeal in a vision of European identity, due to its potential to contribute to the creation of new knowledge, to foster innovation and to engage new companies and research organisations in space activities».[23]

COMMUNICATION FROM THE COMMISSION TO THE COUNCIL AND THE EUROPEAN PARLIAMENT European Space Policy, Brussels, 26/4/2007

The idea that Europe can have a world leadership thanks to her know-how and her technological accomplishments is not new. It seems logical therefore that the Commission should push for an involvement in first-class scientific research in order to promote European interests and leadership, including through space research. It is interesting to notice that the Commission has a double discourse in its communication to the other two European bodies. One focuses on citizens and a feeling of identity, the other focusing on industries/companies and a "de facto" building of an identity through concrete innovation and technological research. An appeal to all layers of society!

2.1.3.10 Exploration as a European instinct

«Cette exploration est [. . .] pour l'Europe la possibilité d'avoir plus de cohésion, plus de sens en se portant vers quelque chose qui représente son identité. L'Europe est conquérante, l'Europe est exploratrice».

«This exploration is for Europe a possibility to increase its cohesiveness, to become a more meaningful project by reaching out for something which represents its identity. Europe conquers, Europe explores»[24]

Claudie Haigneré interviewed by Touteleurope.eu, 18/07/2008

Having been both an astronaut and the French Minister for European Affairs, the current Advisor to ESA Director General knows better than anyone else how important space policy is nowadays for Europe. According to her, exploring and conquering are two intrinsically European attitudes, inseparable from a true European identity. Claudie Haigneré makes the interesting comment that the very act of space exploration is in itself, independently of its results, something representative of Europe's identity. This was valid in the past with the exploration of the Earth, and it continues today with the exploration of outer space.

2.1.3.11 The European Parliament's commitment to ESP

The EP "welcomes the Council conclusions of 26 September 2008 as a useful political commitment towards the development of an ESP (European Space Policy) which contributes strongly to a European identity and reiterates its intention of being constructive and participating fully in its implementation, as if the Treaty of Lisbon were in force"[25]

European Parliament resolution of 20 November 2008 on the European Space Policy: how to bring space down to earth. Point P.1, 20/11/2008

The European Parliament (EP) has always been a strong supporter of an effective EU space policy.[26] It is the EP which asked for the 2003 White paper quoted above after the endorsement of the EU-ESA European Space Strategy of 2001, and in 2004 urged the Commission to support ongoing projects (i.e. to assign a role to the International Space Station (ISS), to become involved in the Soyuz at Kourou project, with a view to ensuring the availability of a manned capsule).[27] EU's unity and identity is one of EP's main concerns, as it is elected to represents the peoples of Europe. Throughout the document quoted, space policy appears as a fundamental tool to enhance such a feeling of cohesion. Since the entry into force of the Lisbon Treaty, Article 189 allows the European Union institutions, including the EP, to play an active role in the shaping of such an identity through space policy.

It seems that from the jigsaw of quotations above comes a varied and yet consistent picture of the importance of space and space policy in building a stronger European identity. Clearly not every author was aware of this, and some quotations might even seem slightly arbitrarily chosen, yet it is paramount to remember that they all contributed in different ways because they appealed to different audiences. Moreover, the "peoples of Europe" are becoming more and more aware of the importance of space in their lives. This is maybe the most solid fundament for strengthening a European identity in space.

---

[1] Landfester, Ulrike. "Laokoon in Outer Space? Towards a Transformative Hermeneutics of Art." Humans in Outer Space – Interdisciplinary Perspectives. Ulrike Landfester, Nina-Louisa Remuss, Kai-Uwe Schrogl, Jean-Claude Worms (eds.). Wien New York: Springer 2010: 159-70.

[2] Translation by the author. Original quotation: «Je suppose, par exemple, que né avec la faculté de penser et de sentir que j'ai présentement, et n'ayant point la forme humaine, je descends du globe de Mars ou de Jupiter. Je peux porter une vue rapide sur tous les siècles, tous les pays, et par conséquent sur toutes les sottises de ce petit globe» Voltaire, *Traité de Métaphysique.* 1734.

[3] For instance, for an analysis of myths related to the constellation of Orion in different cultures, see: 21 November 2010. *http://www.suite101.com/content/myths-for-the-constellation-orion-a26908.*
[4] Youtube. 21 November 2010. *http://www.youtube.com/watch?v=EG7wB3G-xp0.*
[5] Workinghumor.com. 21 November 2010. *http://www.workinghumor.com/quotes/dave_barry_travel.shtml.*
[6] "L'Europe augmente son budget spatial de 20%". 29 November 2008. Le Post. 21 November 2010. *http://www.google.de/imgres?imgurl=http://medias.lepost.fr/ill/2008/11/29/h-20-1341008-1227952360.jpg&imgrefurl=http://www.lepost.fr/article/2008/11/29/1341010_l-europe-spatiale-augmente-son-budget-de-20.html&usg=__L9aqW0vJtm5zz7YuQf0rSQYc0i8=&h=420&w=423&sz=29&hl=en&start=2&zoom=1&um=1&itbs=1&tbnid=YPfod2kmM4vTLM:&tbnh=125&tbnw=126&prev=/images%3Fq%3Ddessin%2Beurope%2Bespace%2Bexploration%26um%3D1%%26hl%3Den%26lr%3Dlang_fr%26client%3Dfirefox-a%26rls%3Dorg.mozilla:de:official%26tbs%3Disch:1,lr:lang_1fr.*
[7] Boston, as recalled personally to the author by a participant to the conference.
[8] Youtube. 21 November 2010. *http://www.youtube.com/watch?v=IRQjjdNL3p8.*
[9] Eurovision website. 21 November 2010. *http://www.eurovision.tv/page/history/thestory.*
[10] Google Image. 21 November 2010. *http://www.google.de/imgres?imgurl=http://www. astronomy2009.org/static/images/stamps/iya_neth.esa.jpg&imgrefurl=http://www.astronomy2009.org/organisation/structure/taskgroups/philately/calendar/apr/&usg=__fTpzQMPjSve-mnoHBdx7WK7nGMk=&h=464&w=591&sz=351&hl=en&start=59&zoom=1&um=1&itbs=1&tbnid=-8MATyEq_AonrM:&tbnh=106&tbnw=135&prev=/images%3Fq%3Dspace%2Bstamps%2Beurope%26start%3D40%26um%3D1%26hl%3Den%26lr%3Dlang_it%26client%3Dfirefox-a%26sa%3DN%26rls%3Dorg.mozilla:de:official%26ndsp%3D20%26tbs%3Disch:1,lr:lang_1it. http://www.space-unit.com/stamps/2009/italy-2009-05-07.jpg.*
[11] Council of Europe. Thirty-Sixth Meeting of the Ministers' Deputies. Emblem of the Council of Europe. Resolution 55(32) of 9 December 1955. Strasbourg: Council of Europe.
[12] "The European Flag". Europa website. 21 November 2010. *http://europa.eu/abc/symbols/emblem/index_en.htm.*
[13] quoted in: J. Krige and A. Russo. A History of the European Space Agency 1958–1987. The Story of ESRO and ELDO, 1958–1973. ESA Doc. SP-1235 of April. 2000. 21 November 2010. *http://www.esa.int/esapub/sp/sp1235/sp1235v1web.pdf.*
[14] Rubbia, Carlo. Edoardo Amaldi, scientific statesman. CERN Yellow Report cernrep 91/09. 21 November 2010. *http://www.amaldi8.org/Binder3.pdf:* 15.
[15] Steven J. Dick and Roger D. Launius (eds.). Societal Impact of Spaceflight. NASA, 2007. 21 November 2010. *http://www.geschkult.fu-berlin.de/e/astrofuturismus/publikationen/Geppert_-_Flights_of_Fancy.pdf: 593.*
[16] "Glass-Steagall Act". Wiki.societal.org. 21 November 2010. *http://wiki.societal.org/tiki-index.php?page=Glass-Steagall+Act.* (Translation by the author).
[17] Gaillard, Florence. "La Construction Symbolique de l'Espace Européen." Hermès 34 (2002): 105–119. (Translation by the author).
[18] Quoted in: de Montluc, Bertrand. "L'environnement International des Prochaines Décisions Européennes sur l'Espace." Politique étrangère 57:1 (1992): 164.
[19] *Ibid.* (Translation by the author).
[20] Quoted in: Steven J. Dick and Roger D. Launius (eds.). Societal Impact of Spaceflight. NASA, 2007. 21 November 2010. *http://www.geschkult.fu-berlin.de/e/astrofuturismus/publikationen/Geppert_-_Flights_of_Fancy.pdf:* 594. (Translation by the author).
[21] "EU approves Galileo Project". 26 March 2002. CNN.com. 21 November 2010. *http://archives.cnn.com/2002/TECH/space/03/26/galileo/index.html.*
[22] Cabal, Christian and Henri Revol. Report: "Space Policy: Daring or Decline. How to make Europe world leader in the space domain". February 2007. Parliamentary Office for Scientific and Technological Assessment. *http://www.senat.fr/opecst/rapport/rapport_politique_spatiale_anglais.pdf:* 11.

[23] Commission of the European Communities. Commission Staff Working Document. European Space Programme – Preliminary Elements. SEC(2007) 504 of 26 April 2007. Brussels: European Union.

[24] "Claudie Haigneré: L'Europe est Conquérante, l'Europe est Exploratrice." 18 July 2008. Touteleurope.eu. 23 November 2010. *http://www.touteleurope.eu/fr/divers/toutes-les-informations/article/afficher/fiche/3611/t/89088/from/4584/arcYear/2008/breve/claudie-haignere-leurope-est-conquerante-leurope-est-exploratrice.html?cHash=6a57ac54e0.* (Translation by the author).

[25] European Parliament. Resolution on the European Space Policy: How to Bring Space Down to Earth. Doc. P6_TA(2008)0564 of 20 November 2008. Brussels: European Union.

[26] See: "European Parliament Fact Sheets. The Aerospace Industry." 21 November 2010. *http://www.europarl.europa.eu/factsheets/4_7_6_en.htm.* the paragraph called "Role of the European Parliament", as well as the forthcoming high-level conference for industry and political decision makers on European space policy which will be hold by the European Parliament in its Hemicycle in Brussels on 26 and 27 October 2010.

[27] Counet, Paul. "European Space Policy. A Shared Competence With Member States." Presentation. Space Weather Week. Nordwijk, The Netherlands. 29 November 2004. 21 November 2010. *http://www.esa-spaceweather.net/spweather/workshops/esww/proc/counet.pdf.* 7.

Education and science communication

2.2 Astronomy and space in curricula – towards a continental education

Peter Habison

2.2.1 Introduction

Astronomy, as one of the oldest sciences, has an innate appeal for people of all ages, partly because it concerns the fascinating, great questions "of life, the Universe and everything" and partly because much of the data obtained with telescopes can be presented as objects of stunning beauty. From a historical perspective, astronomy plays a vital role in the development of natural science and European identity. It were the ancient Greeks who developed astronomy to a highly sophisticated level, European astronomers like Copernicus, Kepler and Galilei who proposed the heliocentric systems and proved the old Ptolemaic system to be wrong and Newton and Einstein who showed the world how gravity functions.

The uniqueness of European science history and the personal fascination of the sky above our heads are key facts when considering communicating astronomy to the public. This native advantage that astronomy has over many other sciences does not, however, relieve us of the obligation to explain what we are doing to the public at large. There are many reasons for doing this. They range from attracting bright young people into the subject to fuel future research endeavours to convincing decision makers to allocate large sums of money to finance increasingly expensive and ambitious projects.

Unfortunately, there are many players out there who obviously haven't gotten the message. Many institutions, agencies, observatories, laboratories and scientists believe that they communicate, but actually they don't. Science writer Dirk Lorenzen from Germany states: "Some of the world's leading observatories only publish a few print-ready pictures per year. Some space agencies operate spacecraft that are virtually unknown to everyone except the most curious enthusiasts for years. These are just two examples of astronomical "communication" of today."[28] For those of you still neglecting science communication, there is time to make a change! On average, scientists and organisations in the U.S. are doing much better in public outreach than their European counterparts. Why is this so? Is it only a matter of funding? In fact, there is a completely different attitude to science communication and most scientists, science organisations and funding agencies in the U.S. have realised that active communication is critically important to keep the system running smoothly and effectively.

For Europe and European organisations it is now time to catch up. This article points out some ideas and concepts with best practice examples from Austria, Germany and other partners in order to strengthen astronomy and space science education and communication, especially in Europe for the coming decade.

2.2.2 Astronomy and its communication to the public

In 2009, professional and amateur astronomers worldwide gathered to celebrate the 400th anniversary of the first use of an astronomical telescope by Galileo Galilei. The launch of the International Year of Astronomy (IYA2009) provided a splendid opportunity to boost worldwide awareness of the subject. UNESCO endorsed IYA2009 and the United Nations proclaimed the year 2009 as the International Year of Astronomy on 20 December 2007. IYA2009 was organised by the International Astronomical Union and reached hundreds of millions of people in over 125 countries. In the final report of this global endeavour, Catherine Cesarsky, former President of the International Astronomical Union, makes the following statement: "Astronomy is one of the oldest fundamental sciences. It continues to make a profound impact on our culture and is a powerful expression of the human intellect. Huge progress has been made in the last few decades. One hundred years ago we barely knew of the existence of our own Milky Way. Today we know that many billions of galaxies make up our Universe and that it originated approximately 13,7 billion years ago. One hundred years ago we had no means of knowing whether there were other solar systems in the Universe. Today we know of more than 400 planets around other stars in our Milky Way and we are moving towards an understanding of how life might have first appeared. One hundred years ago we studied the sky using only optical telescopes and photographic plates. Today we observe the Universe from Earth and from space, from radio waves to gamma rays, using cutting-edge technology. Media and public interest in astronomy have never been higher and major discoveries are front-page news throughout the world."[29]

This statement makes clear that the year 2009 has been a global celebration of astronomy and space science and its contributions to society and culture, with a strong emphasis on education, public engagement and the involvement of young people, with events at national, regional and global levels. It is obvious that its legacy will influence the future of astronomy communication on a planet-wide scale. Now we have to make sure that we gain maximum profit from this situation also for Europe by stimulating the interest and imagination of people of all ages and backgrounds. But how can we do it? Here are a few words about science communication and especially about the importance of astronomy and space science in this context:

Science communication is a complex field, which includes many different disciplines: science outreach, science popularisation, science PR (Public Relations) and even scientific marketing. One of the particular features of science communication work is that it touches upon numerous different topics, issues and areas. Science communication demands knowledge not only of science, but of technology, journalism and of visual communication. Lars Lindberg Christensen writes: "For any branch of science it is necessary to find its communication niche. As an example, communication of astronomy and related (space) sciences are just sub-branches of the more general field of physics communication or, even more broadly, communication of natural sciences. Astronomy does however, play a special role in the field of science communication. It covers a very broad area of research with instant photogenic appeal and a scale and scope that go far beyond our daily lives to stimulate the imagination."[30]

And he concludes: "On top of all this, astronomy touches on some of the largest philosophical questions of the human race. Questions that seek to explain our very existence. Where do we come from? Where will we end? How did life arise? Is there life elsewhere in the Universe? This, and more, gives astronomy special benefits in the 'battle to be heard', and, to some degree, astronomical institutions are using the appeal of their science more extensively than many other branches of science. Also, since astronomy, in most respects, has almost no direct practical application whatsoever, the need to excite the population with good results is even more important than in other branches of science and so, possibly, astronomical institutions are just one step ahead in the science communication game. In summary, astronomy can lead the way for other natural sciences and be a frontrunner in science communication. Astronomy has a natural ability to fascinate and can open young people's minds to the beauty of science. There is, however, a big "but": astronomy is also, practically speaking, fairly useless and applicable results from this kind of fundamental science can take centuries to materialise. We should make this clear to ourselves and answer questions from the media about this issue honestly and play on the "inspiration-factor" of astronomy and the general value of fundamental research instead."[31]

2.2.3 Astronomy education and best practice examples

It is this inspirational factor and innate curiosity about the world in which we live, that draws people towards astronomy and space, providing rich opportunities for outreach and education. Planetaria, Observatories and science centres are the natural conduits through which the flow of astronomical information is disseminated to the wider public. The European space-related Agencies (ESA/ESO) have worked in collaboration with some of the major planetarium associations in Europe, but to increase the impact of European astronomy communication to society, a more systematic collaboration and coherent strategy is strongly required.

Here we present a collection of five "best practice examples" in Austria, Germany, the U.S. and throughout Europe in the years 2007 to 2010, all dealing with space science and astronomy communication.

2.2.3.1 Education project „Der Himmel für Schulen"

In the context the International Year of Astronomy 2009 Vienna Planetarium and Observatories developed an outreach project for schools in collaboration with the Austrian Federal Ministry for Education, Arts and Culture, the Federal Ministry for Transport, Innovation and Technology as well as the Austrian Research Promotion Agency. The project ran from February 2009 to January 2010 and focused on three major topics:

- Analysis of astronomy education in primary and secondary schools in Austria as well as opinion polling at Vienna Planetarium and Observatories with the help of questionnaires and direct inquiry. Subject to the questionnaires and interviews were for the greater part teachers and students, but also experts in the fields of presentation techniques, graphics, astronomy and pedagogics.
- Planning and conducting of pilot projects in the fields of astronomy and space science with primary and secondary schools in Vienna and Lower Austria.
- Development of a strategic plan for a new Austrian astronomy education concept in order to strengthen communicating astronomy to schools and the public.

Based on the results of the opinion polling among teachers and students as well as the lessons learned from the conducted pilot projects the authors[32] recommend the following strategies:

Any actions taken should concentrate on schools of general education and especially primary schools, for they would reach the majority (70%) of Austrian school students. Primary schools should furthermore be the focus of any measures taken, because enthusiasm and interest in astronomy as well as sciences and engineering in general should be encouraged at the youngest age possible. The obtained data shows a strong trend of young women towards higher education in general and science in particular. Still their share in mathematics, astronomy, physics or engineering and mechanics is rather small and measures should be taken to encourage young women to follow an education and career in the above-mentioned fields.

The attained data reveals the lively interest students and teachers show to astronomy and spaceflight. This enthusiasm can be encouraged and increased by a large number of very diverse outreach activities. Still it is advisable to seek out the most promising focus group among the different types of schools and educational levels. Comprehensive education in primary schools combined with a large number of available day trips make these schools a highly attractive subject for outreach activities to further astronomy education at a very early stage. Although primary school teachers are not bound to cover a lot of astronomical topics as far as their curriculum is concerned, they still have the benefit of flexible lessons, sufficient abundance of day trips and usually very interested and enthusiastic students.

Another focus should be put on physics teachers and their students in secondary schools. Even though physics is usually allotted only one to two lessons per week, it is still the subject with the largest abundance of astronomical or related topics. Education offers for physics teachers also have to take into account the fact that

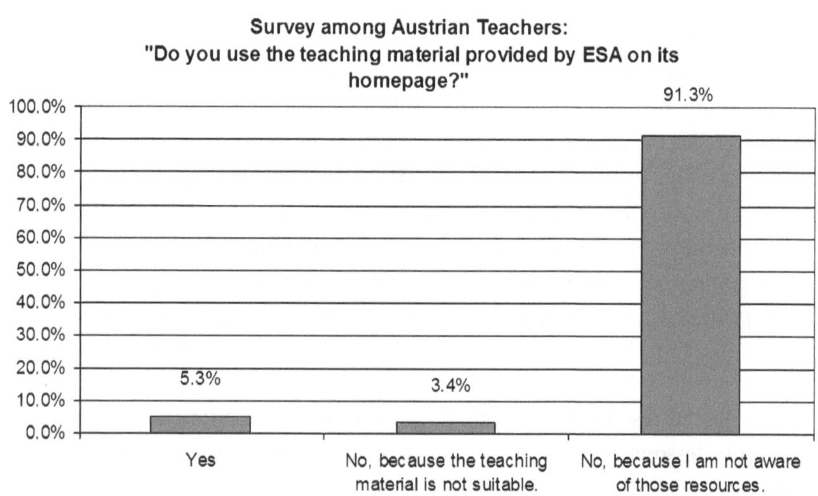

Fig. 2.1. *Survey among 147 Austrian teachers: "Do you use the teaching material made available by ESA on its homepage?"*

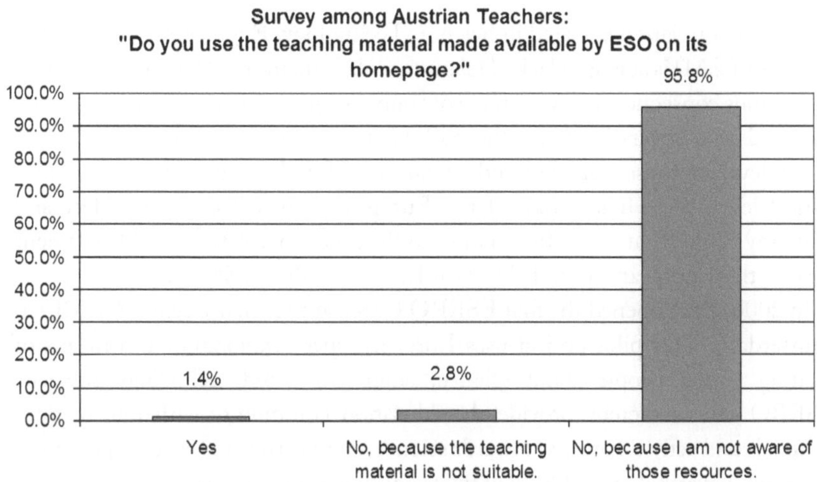

Fig. 2.2. *Survey among 147 Austrian teachers: "Do you use the teaching material made available by ESO on its homepage?"*

they not only have little time to spare – regarding the number of physics lessons per week – but also have to share the available day trips with their colleagues.

90% of the surveyed Austrian teachers are not aware of the wide range of education resources provided by ESA or ESO. Therefore ESA as well as ESO should seek to heighten their level of awareness by establishing permanent co-operations or representations in Austria. ESA has already instituted European Space Education Resource Offices (ESERO) in several Member States and could increase the abundance of astronomical education in Austrian schools and most importantly further the motivation of young people to pursue an education and career in the field of astronomy in particular and science, technology and engineering in general by implementing a European Space Education Resource Office in Austria.

2.2.3.2 European Space Education Resource Office

Through the ESERO offices ESA's Education Office aims to support the specific educational needs of the Member States and their education community and to get easy access to national education networks. The decrease in students taking up classical science and engineering subjects represents a real problem for organisations like ESA. The recent interest in applied sciences such as information and communications technology and biotechnology has not been enough to offset the downward trend, especially among European women. As the current

European population engaged in Science, Technology, Engineering and Mathematics (STEM) is ageing, a lack of talent to replace them could have wide-ranging and serious consequences. A shortage of young people going to university to study these subjects adversely affects the STEM industry and employment markets. If the level of those qualified and literate in STEM subjects drops beyond a critical level, the whole future of the European knowledge-based society and economy could be at stake. In response to this, the ESA Education Department created the European Space Education Resource Office (ESERO).

In 2006, ESA opened the first ESERO Office at the science centre NEMO in Amsterdam. This pilot project uses European space exploration as a means of exciting young people about science, engineering and technology subjects. ESERO offices thereby provide the education community with information, materials and activities geared towards science, engineering and space exploration. There are nevertheless problems with attempting to provide this kind of support: reaching the millions of students and educators throughout Europe is an impossible task for ESA. In addition, there are all the different languages, as well as unique education systems in each Member State. Therefore ESA decided to pursue a state by state approach, which fosters close ties to education stakeholders. The ESERO office tailors activities specifically to the students of the Member States and it uses space related themes and the genuine fascination for spaceflight to present science and technology to young people. It will promote science and engineering, careers in the space sector and provide support for the delivery of national curricula. It will be therefore manned by an expert, well integrated into the local education system and networks. This would allow ESA Education to support the individual needs of the Member State, while simultaneously having access to those national networks of publishers, museums, teachers associations etc.

The first implementation phase began with Belgium, the Netherlands and Spain, each having an ESERO office. The first contract was signed in the Netherlands in late 2005 with NEMO in Amsterdam. The ESERO office in the Netherlands delivers information and support to educators by developing teacher (primary and secondary) training sessions in collaboration with national partners. These training sessions consist of interactive lectures and hands-on workshops related to different space themes, like satellites, weather and seasons or the lifecycle of a star. ESERO disseminates existing ESA education materials, and if appropriate, develops specific resources tailored to the needs of the education community. The ESERO office also organises annual ESERO teacher conferences for secondary and primary education.

A good example of what ESERO Netherlands accomplished is the publication of a complete book with 80 lessons about space for primary education, which fits well into the Dutch school curriculum. The book is called "Travel through space in

80 lessons". The book was developed because of the huge demand for good educational material about space. The book focuses on all groups of Primary Education (age 4–12). The content has been based on the pedagogical philosophy "Inquiry-based learning and learning by design". For support, ESERO developed workshops to give teachers hands-on insight in the use of this book. For Austria an ESERO strategy will be developed in 2011 and the opening of the Austrian office is planned for the year 2012.

2.2.3.3 400 Years of European astronomy – collaboration between ESA and Europe's planetaria for the IYA 2009

Most of our knowledge about the Universe is based on findings made by European astronomers like Kepler, Galilei, Cassini, Newton, Herschel, Planck, not to mention Aristotle, Ptolemy, Archimedes and numerous other scholars and scientists of the past. Europe's heritage of almost three millenniums of astronomical observation and scientific research has formed a matchless scientific community. *Astronomy is part of European identity*, coming down from world-class institutes to hundreds of thousands of amateurs and interested laypersons.

Despite spectacular findings and breakthroughs, the significance of European astronomy is shifting off the focus in general public – especially space astronomy "made in Europe" is lacking public awareness: the Hubble Telescope with its huge impact in celestial imagery is hardly identified as a project with a 15% European funding and even larger involvement. On the occasion of the International Year of Astronomy 2009, ESA was producing for the first time a large, high quality planetarium show, dealing with space astronomy and its stunning results. The story begins with Galileo Galilei, who 400 years ago used a telescope for the first time, and has changed the way in which we see the Universe. Galilei has paved the way for our modern scientific understanding, resulting in space telescopes centuries later to scan the sky in all wavelengths, including the Herschel and Planck cornerstone missions of the agency.

The show was produced in full dome video format with real video, 4K quality animations and space imagery shown for the first time. In parallel, a classical version for all planetaria with analogue equipment was prepared, as they still represent the majority of planetaria in Europe. The story has a European touch: didactic, educational and still emotional and human. The show was released in spring 2009.

For the first time a new cooperation model between planetaria and the European Space Agency was put in place. Initiated in Austria, Germany and Switzerland,

Fig. 2.3. *Official poster of the ESA 2009 planetarium show "Touching the Edge of the Universe".*

a pool of interested planetaria and ESA elaborated a concept and roadmap on how to create a product, being similarly useful and fitting for planetaria purposes and supportive to Europe's contribution to space astronomy. The result was more than promising. Most of the large planetaria in the German speaking countries have signed an agreement and contributed to the production. Other theatres followed in 2009/10 and the planetarium show was adapted to other European languages and brought to several planetaria of the world. The programme with its emotional and highly visual content contributed to promote European space astronomy within the general public.

2.2.3.4 ESO Science Outreach Network (ESON)

The ESO education and Public Outreach Department (ePOD) has established the ESO Science Outreach Network (ESON), a network of persons in ESO Member States and other States, who serve as local contacts for the media and general public in connection with ESO developments, press releases, exhibitions, etc. At the same time, they also provide contacts between the media and the scientists in their area. The ESON members are typically full-time science communicators and know the field and its movements. They know the national players (media, academia etc.) and regularly interact with them, are able to find a national angle and provide regular inputs and ideas on how to best reach the target groups in their area. Some objectives and deliverables of the ESON members are: translation, adaptation, distribution and promotion of the ESO press releases, showing the national benefit of ESO membership (scientific, industrial), being a point of contact for the national press, integrating ESO into existing outreach efforts, providing regular updates for the ESO national WebPages and promoting ESO in public events and educational activities.

ESON and the ESO Educational Office could be the starting point to build a tight cooperation between ESO and the education community. ESO develops the concept of "Partner organisations", where each partner organisation and ESO benefits from each other through their collaborations on a bilateral basis.

2.2.3.5 The Digital Universe 3D Atlas and Uniview[33]

UNIVIEW is a visualisation and simulation software application aimed at presenting and teaching astronomy, astrophysics and other sciences. The programme provides a seamless visualisation of the entire known universe, from human scales all the way to the cosmic horizon. It allows immersing the audience in fascinating stories and creates unforgettable experiences for visitors of all ages. It comes with the world's most comprehensive three-dimensional astronomical database – the Digital Universe from the American Museum of Natural History (AMNH). In 1997 AMNH was funded by NASA to develop an evolutionary recasting of planetarium presentation based on immersive 3D data visualisation of the current understanding of our astronomical surroundings. The Digital Universe, a compendium atlas of all known academic astronomical catalogs published with distance information was a direct result of this partnership as the old Hayden Planetarium, housed at AMNH, was rebuilt to accommodate full dome, high resolution digital projection. Surround visualisation of the Digital Universe (DU) became the basis of a new breed of planetarium shows based on true data

Fig. 2.4. *The inner solar system with Mercury, Venus, Earth and Mars; UNIVIEW, Carter Emmart.*

visualisation. The interactive viewer for DU, called UNIVIEW grew out of these efforts.

From the start of assessing how best to present DU to future planetarium audiences, it was observed that linking multiple users into the same shared virtual world space was beneficial. The success of connecting the NCSA CAVE with the new Hayden Planetarium was multiple: remote decision making groups would examine the same data in shared virtual worlds. This feature was designed into Uniview in the very first stages of it as a thesis project internship at AMNH. Unview's "Octopus" feature has enabled a global reach between users for demonstration and teaching. The concept is to bring remote authoritative guidance and description of the data being displayed in DU to distributed audiences and classrooms across the world. Control requests also enable participation from audiences to take place in an orderly fashion.

Today, we have the technology to integrate the mapping of what we know about the universe and share it among the citizens of the planet. The need to organise the approach to teaching and professional development with this broad palette will likely emerge with the help of those who know its components best and can share that with the rest of us. Archival of these narratives and guidance will likely serve us well as we move further into these tools which help us understand an ever increasing vast inventory of information.

2.2.4 How do we communicate? – forging a path to the future

Talking about media, communication and networks, what will the future bring, especially for science communicators? In his paper "Where is everyone?",[34] which is based on a year-long gathering of anecdotal evidence, Thomas Beakdal, a writer and social media advocate, tries to answer some fundamental questions and he writes: "These days everyone is trying to connect with other people. It used to be simple, but in these technological times it is a little more complicated! We are currently in the midst of the most drastic change since the invention of the newspaper." Talking about the future we have to answer questions like "how do we connect with other people today?", and more importantly, "how will we do it tomorrow?" Here comes a short summary of Beakdal's findings:

He starts by looking back to the year 1800, when finding out information was very difficult. The only way you could really interact with other people was to go out and meet them; it was all about face-to-face communication, best at the market-place of your local town. In order to receive or give information to other people you had to be right there, you wouldn't know what was happening in another part of the town or even in another city. By the year 1900 newspapers and magazines had revolutionised our communication; now people could read news from places where they have never been and could communicate their ideas to people they had never seen. This was the first real information revolution; the world was opening up to everyone.

From the 1920s on a new source of information caught people's attention: the radio. Now people could listen to another person's voice hundreds of miles away. But most importantly, they could get the latest information live. It was another tremendous evolutionary step in the history of information. During the next 40 years the next technical revolution, television was introduced. It began to gain public interest in the 1950s, and by the early 1990s its presence was huge, effectively surpassing newspapers and magazines whilst dominating the radio. Now people could both hear and see information.

In the mid 1990s a new phenomenon loomed in the shadows: the internet. 1998 was the year when the internet changed from being a place that had little relevance to being arguably essential. Every company started to think about their need to have a website and the internet philosophy caught everyone's attention. Nevertheless it was relatively little-used and most people did not have access to it in those days. People also started to realise that the internet was more than just information. You could give feedback by joining the conversation and being part of the experience instead of just being a spectator.

In 2004, only six years later, the internet had revolutionised how we approached information. Television and newspapers still dominated our news sources, but the new world was definitely online. In the same year another new phenomenon started to take off: social networks. The concept had been slowly gaining ground with the concept of blogs. Three years later the social element of the internet showed just how powerful the voice of the people really is. Everyone wanted to create their own little world and connect it to their friends. Thomas Beakdal states that the year 2007 was the turning point for traditional websites, people wanted active information instead of a static and passive form of information.

In 2010 the new internet dominated our world completely. Traditional websites are dying, killed by the constant streaming information from social networks. This concept is the most dramatic change and is quickly taking over our need to stay up to date with what goes on in the world. News is no longer being reported by journalists only, as now it comes from anyone and everyone. It is being reported directly from the source to you – bypassing the traditional media channels.

Nowadays a new concept in the form of targeted information is slowly emerging. We are already seeing an increasing number of services on mobile phones that provide local information for the area that you are in. Hundred of thousands of application (apps) are already available and this is something that will explode in the years to come. In a world where we have access to more information that we can consume, getting the relevant information is going to be a very important element.

2.2.5 Recommendations and conclusions

In this context, it is certainly an important momentum for astronomy communicators, with numerous opportunities to promote how science can have a positive influence on society. It is vital that the organisations providing astronomical facilities and also individual scientists recognise the importance of explaining what they do to the people. The general public are ultimately those, who are paying them to do it. By ensuring that public communication is seen as an integral part of a scientist's job and that it is given clear recognition when done well, a culture of high quality communication can be encouraged.

Here are some recommendations concerning strategic planning for science communication and public outreach for the coming decade:[35]

1. Strategic long term support for public communication and education in Europe must be provided.
2. Active steps should be taken to strengthen the links between science museums/ planetaria and the European Agencies (e.g. ESA and ESO).
3. Provide media training and science communication courses for scientists and encourage them to utilise these courses.
4. Provide a standardised science communications portal for media, educators and interested laypeople.

5. Communication departments must be organised and operated in a professional fashion – science communicators working together with active scientists.
6. Forge collaborations with science communicators in the U.S. and overseas.

These recommendations have also to be seen in the framework of European identity, reflected in European education science programmes. We know that there is still a lack of awareness among European citizens on European astronomy programmes and space endeavours, but nevertheless: Europe is back on track. Big astronomy and space programmes in the last decades (e.g., ESO-VLT, HIPPARCOS, ISO, Huygens, Mars Express and most recently Herschel, Planck and E-ELT) contributed to the visibility of Europe in the world and allowed European citizens to identify themselves with such projects and space science in general. This reinforced Europe's sense of common identity and uniqueness in the world and helps us to play in the first league of space nations.

However, we have to think even further. Our vision must be to help people rediscover their place in the universe through the sky, and thereby engage a personal sense of wonder and discovery. Everyone should realise the impact of astronomy, space science and other fundamental sciences on our daily lives, and understand how scientific knowledge can contribute to a more equitable and peaceful society. In 1939 Antoine de Saint Exupéry wrote the following lines:

> "To be in love is not to look at each other,
> But to look together in the same direction"

This quote also holds for our "love and passion for astronomy". So we want to look together into a bright future of "Communicating Astronomy and Space with the Public" – a subject that is more important than ever before and whose significance is growing nearly every day.

[28]Lorenzen, Dirk. Foreword in: Lindberg Christensen, Lars. The Hands-On Guide for Science Communicators. Springer, 2007: vii.
[29]Cesarsky, C. The International Year of Astronomy 2009 Final Report. International Astronomical Union. 2010: 17.
[30]Lindberg Christensen, Lars. The Hands-On Guide for Science Communicators. Springer, 2007: 5–6.
[31]*Ibid.*
[32]Fischer, Monika, Habison Peter. Final Report of the Education Konzept„Der Himmel für Schulen". Vienna, February 2010.
[33]Based on the paper "The Digital Universe 3D Atlas and Uniview – Rethinking the planetarium concept" from Carter Emmart in „Himmel@All – Astronomie in Bildung und Kultur" 2010, Vienna.
[34]Beakdal, Thomas. "Where is Everyone?" CAP Journal 6 (June 2009): 13–17.
[35]Ros, Rosa M. et. al., "ASTRONET: Public Outreach." CAP Journal 5 (January 2009): 26–31.

Anthropology and philosophy

2.3 Europe, peace, space[36]

Moniel Verhoeven in cooperation with Ignatius van Neerven

2.3.1 Introduction

During the night of 25 October 732, the Islamic army of Al-Andalus leaves the banks of the river Vienne near Poitiers and retreats to the Pyrenees. The Frankish warriors have killed its commander-in-chief, Abd al-Rahman in one of the clashes some days before. The Francs pursue the defeated army and batter it like a hammer on an anvil – hence the name given to their leader, Charles "the Hammer". Some years later, a monk in Cordoba writes the story of this war. For the first time in history the concept of "Europeans" appears: *europenses*. The anonymous author implies in his text that Europe constructs itself as a reaction to Islam.[37]

Has Europe indeed constructed itself as a reaction to Islam? Was Europe not born on the Agora in Athens, in the Senate of Rome, or in the coffeehouses of Vienna? What does it mean to speak of the concept *Europe*? Does it mean anything at all?

In the first section, I will be searching for an answer to the question: What does it mean to say "Europe"? In the second section, I will consider the link between the *idea* of "Europe" and the construction of the European Union. In the last paragraph, I suggest that "Europe" has a specific vocation to explore space.

2.3.2 Europe: rupture, transport, transcendence

> *"Europe is a genesis whose contours are not to be traced out beforehand or even to be predicted precisely"*.[38]

Libraries are filled with books about the "essence" of Europe, her "identity", her "core", her "spirit". The philosopher Denis Guénoun tries to avoid this essentialist trap. He is not in search of a fixed identity or an essence.[39] Instead, he searches for the *figure* of what is named "Europe". A figure does not designate anything existing, but the beginning of a "becoming-shape". In his book *Hypothèses sur l'Europe,*[40] he defines the specificities of what we name "Europe". His central thesis is that Europe is not a foundation, nor a fatherland nor an objective. It is not a territory: Europe is an idea, an idea of a process. This idea is born out of a specific ambivalence and even misalliance: the crossing between a desire for universalism and the concern for its repatriation. The content of this idea is universalism:

> *When you search the idea of Europe, you find universality. The positive content of the culture, which has been elaborated in what we call "Europe" consists in nothing else but the determination, the digging out, the exploration of the most common and the most shared: intelligible unity of the natural world, common condition of human beings. This approach is not initially acquired; instead, it is a process of continuous expansion: the universal is gained intellectually and in practice by an extension towards new physical spaces (. . .). The universal is a universalisation, a march in progress. Europe is a place and a road towards it.*[41]

I will follow some stages on this road.

2.3.2.1 Athens: rupture

Europe has no origin, no place of birth. The maiden Europè[42] has been raped from the coasts of Asia and her name became the name of the continent we speak of.[43] The myth of the princess Europè is a story of transmission: the "origin" is a transfer. That is the originality of this origin. Herodotus uses the word "Europe" for the first time. In his work, "Europe" does not designate a place on the map, but a relationship of places, a passage, a voyage.[44] Europe should be thought of as a median space, intermediate, maritime, a threshold of crossings and bridges, from where one sees the land, the rivers and their relationships. Europe should be thought of as a moment, as an era, as a traversal.[45] Traverse it, to leave it. Europe is the *passage* to Europe.

Europe is not the name of a piece of the Earth, but the name of abduction from the known place of birth towards a place in the unknown and unnamed West.[46] It is about a desire to go beyond the horizons, a desire to uproot the heritage, a renouncement of autochthony and, consequentially, access to the freedom of the universal dimension of the world. Europe is the return of the universal to itself. The "universal" conceptually precedes Europe, but does not require it. The movement of the universal was brought to Europe. The Greeks did not invent the universal: they transmitted it. Since the first Greek transfer that inaugurated the chain of recitals and heritages, the "universal" has expanded to just about the whole world. The name for this movement of expansion is: history.

The Greeks did not consider themselves Europeans: they thought and spoke of themselves as *Hellenes.* However, they were aware of what they were: the Greek opening is an opening towards infinity. In inventing philosophy, they created the possibility of a new type of humanity, which is in fact humanity itself, open to the infinity of its own. It is the search for truth, and the source of truth is in the individual, in the soul. It is the relation of man to himself and to truth, to him as being open for truth: this is the foundation of justice. Justice is not between men, but *in* man. This interiority is the possibility, the foundation of politics: what men have in common in the context of politics is not mutual respect, but respect for themselves. What men have in common is their interiority. Politics is the sharing of a community of intention, the direction of which is determined in the privacy of intimacy in which the individual constitutes, by himself and he himself alone, the guarantor. Since care for intimacy, care for the soul (Plato) is possible, the community and the State are possible.[47] *Ecclesia* is the assembly of the people that constitutes and governs the *polis* as an entity. Freedom is guaranteed for the citizens of the polis; women, slaves and *barbaroi* are excluded.

The period of the assembly produces another assembly: the theatre, more exactly, tragedy and comedy.[48] Theatre is the putting of becoming "common" into play by giving the members of the assembly the capability to verify how everyone is the unique trustee of concern for the soul and for the truth, and the foundation of the political community. In the theatre, the world, as told and commentated, does not represent itself. The *cosmos* can enclose a *we* that interpolates: an assembly, a theatre. What assembles is the polis: the polis is an assembly: it is a space of *us*: the Citizens, the Athenians. In the polis there is no mere representation of the world; there is a *cosmos.*

2.3.2.2 Rome I: transition

Rome follows Athens and prefigures Europe. Rome is transport. This prefiguration has two names. The first of its names is *Empire.* The second name is *Ecclesia.* These two names designate the transition of Europe to itself and beyond itself.

Rome invents a non-urban citizenship: one can be Roman everywhere. Metropolis and colonies are equal in citizenship. However, one difference stays: the new City, the political corps dispersed in citizens in all the corners of the world, cannot reassemble.[49] There is no assembly of the Republic. Politics is *we.* With ongoing colonisation, something else is also produced: the world, the *mundus* that is no more the Greek *cosmos*. The idea of *mundus* means first that the world should be thought of as something "in front". The world is not "around": it is "outside" myself and is put in front of me. The constitution of the world as world, designates precisely the process whereby the Republic breaks up and gives space to the Empire. The Empire is necessarily the empire of the world: *imperium mundi*.

In the Empire, there can be no assembly. So Rome produces the image. The world must be *imagined*, presented in a spectacle: it is the system of images of the world, because the world is that which can only be represented. The world is submitted to the authority of the Empire, is constituted as a world by the exclusion of the possibility of an assembly and thus of becoming "common". The world, without the power to re-assemble itself, can only be convoked in a direct frontal confrontation, the spectacle. The world is that of which there is an image: *imago mundi*.

The universal as transport is the "essence" of *romanity*: Rome "invents" the universal. This is why the central space of Rome, its "interior" is the sea: a kind of "pure" transport. Water is the element of movement, of mobility. Every foundation is "earthly" by essence. The maritime element does not found anything; it carries, it transports and it transfers.[50]

The specific Roman invention is the law. Not the idea of justice, or of a process, or of judgment, but of what we can call juridical: the regulation of transactions and of transfers of transports and exchanges. The law is the ab-straction of the universal.

2.3.2.3 Rome II: transcendence

The establishment of the Empire of the World, Rome I, is based on the impossibility of the assembly. The assemblage is constituted as otherness that escapes and transcends the Empire. Since there cannot be an assembly, there cannot be any more care for the soul. Therefore, another *ecclesia* assembles: it installs itself as the *assembly of the people of God, Ecclesia*, Church. The Church takes care of the souls, but in a transcendent dimension, indifferent to the actual Empire. The assembly of the people of God is born as a *new ecclesia*. Therefore, the Church is the return towards itself from the universal figured as an assembly, in the *regime of the assembly*. Sovereignty of the world and assembly of the people are formed in this way as rival and separated regimes. The Empire and the Church are twin productions, but inversed: sovereignty without an assembly; an assembly without power. There is an articulation, a rupture between the Emperor and the Pope: a separation between sovereignty and the assembly.

The Empire has a seat: Rome. The Church also has a seat: Rome: this is the *romanity* of the world. Rome is the proper name of a common place. The theoretical name of this mimetic and infinite ambiguity is the theological-political.

2.3.2.4 Europe

For the moment, we have two figures for the universal and they are tied up together: the Empire and the Church. A third, Europe, succeeds them. This transition starts with an idea. In the heart of the Middle Ages arises the fiction of the Holy Empire.[51] This Empire wanted to be the reunion between the North and the South, a synthesis between *romanity* and *germanity*. The idea carried the name of these four elements: *Holy German Roman Empire*. Nevertheless, this synthesis would become a failure: it remained a political-religious fiction.

Why did this effort not succeed? Europe did not return the universal as Empire, nor as a Church, nor sovereignty or assembly. Europe is neither a kingdom, nor a people. From the Roman prefiguration, Europe takes the possibility of *continentality:* it takes the shape of a continent, an earthly and human reality. It had to, as the transport of Europe is enclosed: in the West by the Atlantic Ocean, in the North by the Arctic Sea, in the South by the Mediterranean and most of all, by Islam. Islam had tried to approach *romanity*, integration in the *mundus*-relic of the Roman Empire, but renounced this transport. It dedicated itself to the absoluteness of a place. No transport anymore, but return. The Islam *is*: there is no concept such as *becoming*-Islam. Europe became a *continent* facing Islam. Therefore, for the first time, the anonymous monk of Cordoba wrote *europeensis*.[52]

This shaping to *continentality* has several consequences: Europe becomes the figure of a return to the earth, as a "folding" on its *continentality*, but also as a return to an origin, that is a production of origin: origin of the universal, foundation, initial rise of the idea of universality. Europe becomes the earthly figure of a fatherland, the fatherland of the universal. Europe is the giving up of each foundation.

Europe is always a "project of the world"; the production of globalisation. The *continentality* of Europe is its return "to its own", its withdrawal. Europe reconsiders the globalising movement that she carries and she re-thinks it by going back to its starting-point. Expansion and "withdrawal", transport "to go" versus transport "to return", metaphoricity-figurality: these concepts construct the concept of Europe.[53] Europe is a "median", intermediate object. Europe is a passage, a process, a *mi-lieu*. Europe takes passengers on board. However, Europe is a movement, without an origin, and a return towards the origin of the movement. This schematisation invites thinking Europe in a double way: first in the movement of universalism (as a moment, par course, trace of this movement), second in an inseparable way, in the withdrawal of this drive, its enclosure. This double movement of withdrawal and renaissances leads to an "open space" of creativity and innovation.[54]

2.3.3 Peace

> *"If we are searching for an answer to the immense silence of the universe, we only can find it in our own responsibility"*[55]

The last of the great silent movies is about space.[56] Fritz Lang directed in 1929 *Frau im Mond.* In this movie, the protagonists Helius and Friede (mark their names!) express their love. They stay on the Moon, while the two other survivors of the expedition return to the Earth: the Moon has been conquered; love arises after hardship and betrayal. Forty years after *Frau im Mond*, Stanley Kubrick produced *2001: a Space Odyssey.*[57] This movie gives us a similar ending: a message of hope, hope for a *new man*. The child, the *new man* seems to be born out of the exploration of space.[58]

Both movies express, on an imaginary level, some human dynamics inspired by the discovery of space. Hope for the blooming of love, by accepting human difference and the desire of the birth of a "new man" form the ends of these two landmarks of Euro-American art. Both movies symbolise the link between the exploration of space and the idea of peace. Is there indeed such a link? Is the striving of Europe to universalism leading to a specific approach to space research? However, how do we perceive and represent "space": does it have to do with the definition of borders, with ways to co-exist, with hope, or is space only there as a neutral playground for our technological discoveries?

2.3.3.1 Ruptures and renaissances

Europe is a privileged space for human expectation. She resurrected out of ruptures and conflicts. These conflicts are in a certain sense even her internal force. They have been constantly reinterpreted and followed by renaissances, accompanied by the movements of hope and pardon. Europe comes from "somewhere else" and moves elsewhere. We saw that Europe started over again out of a rupture, out of a "founding lack". From a mythical point of view we saw that Europe has been "raped", "torn off". We saw that this consciousness was also the beginning of a new movement. Europe incarnated the idea that it is "moving". It is a construction "en route", "on the road". Europe dares to transgress her own frontiers and limits, always by a double movement: an effort to "repair" structures, and a movement of hope and revelation to come to "something new". Her lack of origin, her uncertainties and "not-knowing" are also forces to go further.

Immanuel Kant[59] elaborated this idea of transgression as an orientation towards "open universalism". The "identity" of Europe is universal, it is the cradle of cosmopolitanism and in fact, a transcendental idea, namely that it is a territory without clearly defined limits. The idea of Europe is to create "eternal peace". The way to develop this peace is: national states should be organised on a federal level, cultural diversified international laws should be developed and a cosmo-political world citizenship should be realised. Kant states that the European commercial spirit is a means to achieve this purpose, provided it is based on mutual respect.

The source and purpose of the European project is to create peace. The main question is how to achieve this: is a juridical approach with an external legalism enough, or do we have to develop our interior life, our consciousness to create peace with our demons? Maybe the force of this thinking is embedded in the fact that Europe does not pretend to have roots, or to have been founded. On the contrary, Europe is a hope that starts somewhere "in the middle": neither the beginning nor the end is essential, but the idea of "mi-lieu" carries the concept of Europe.

"Space"[60] is thus an essential figure of European civilisation. Europe is consolidated on a basis that transgresses frontiers: common values. In addition, Europe has transformed these common values into universal values. Some of these are: the pluralism of ideas and beliefs, intellectual curiosity, the spirit of revolt and the negation of fatality, the perfect combination of spirit and will, the taste for adventure and action, life understood as a defiance and science as the expression of the human will to control nature and utilise it for its needs, and liberty, in the sense of a metaphysical concept as well in the sense of political principle: the civil, "earthly" liberties.[61]

Specific to the European tradition is the concept of renaissances. The European constitution is one of reinterpretation and redefinition towards its own traditions. The *roman* European realises himself that he cannot become an owner of what he did not possess, or go back to an origin that never existed and that he cannot be completely unified with his own tradition. This is why European culture can be characterised by a consciousness of transcendence, both in its scientific design and its political and cultural organisation. Europe is a double movement: evolving to infinity and redefining limits and frontiers. Europe is by definition un-accomplished.

2.3.3.2 A political renaissance

The Treaty of Rome signified the triumph of a very realistic, gradualist approach to build a new political Europe. Jean Monnet inspired this method: structuring Europe by small but irreversible steps. Monnet, Schumann and De Gasperi were clear about the motive of developing a European construction: peace through cooperation. The Treaty of Rome hoped to "provide **peace** and prosperity by establishing the foundations of an ever-closer union among the peoples of Europe."

Developing a European space programme was born out of the same idea. The second article of the Convention of the European Space Agency (1975) states this very clearly: "desiring to pursue and to strengthen European cooperation for exclusively **peaceful** purposes".[62] The construction of Europe can be seen as an operation "sui generis", without a precedent and, for the moment, without a successor. Europe has the courage to be constructed on its political and economical disasters.

The governance of decision-making is unique. We call this the communitisation of decisions. The exercise of citizenship implies that European citizens everywhere are eligible, actively and passively, in local and European elections. Dissociation between nationality and citizenship has been realised. Citizenship has been detached from the nation-state.[63] Territory is not nation: citizenship transgresses the state.[64] European space[65] is transnational and does not signify a policy of territory.

In addition, there is international law and public law. More or less 80% of the laws that govern the daily lives of individuals in the European Union are European. The Court of Justice has become a power over all the territory[66]: it is an indirect influence. Europe is becoming a subject and becomes less an object: we are directly implicated in a development that transgresses the nations.[67]

Concerning the relationship between space and identity, a consciousness has progressively grown in Europe that an "identity" is not unique, but that a plurality of "identities" exists. Thanks to Europe, one can see the plurality of "belonging".

The European desire for universalism and globalisation has created a planetary area and has introduced European technology and its organisation.[68] According to Jan Patočka[69] the principal problem for Europe and her nations is the fact that biological-technical rationality dominates. Edmund Husserl has cautioned already of separating factual scientific knowledge from the subjective instance that knows. When a society is only oriented towards the knowledge of facts, the risk of a non-critical community arises.[70]

Jan Patočka wonders how Europe can develop new answers for this relationship between the bio-technological developments, its organisation and the concentration of power. He wonders if Europe, by exploring her own limits, can produce a horizon of new and authentic possibilities. Philosophy has already been replaced by factual science; what we now see is "a frenetic technological development, which trims the pathways of social realisations of great extent". With the progress of the nation-states in Europe, science and technology have been distributed to them according to their power and respective superiority.[71] Europe is vulnerable towards the question of the "common element": how can the technological and commercial monopoly that Europe has developed become organised in what has been defined as "the world"?[72] In other words, can a European "spirit" be envisaged that creates a new form of hegemony, "rejuvenated and enriched by the modern theme of liberation"?[73]

2.3.3.3 Globalisation and peace

The process of globalisation leads to a new consciousness of "limitless". It evokes the sentiment of a "cosmos" into which man is thrown and somehow lost. This is the universe as explored by astronomers, astrophysicists and astronauts.[74] The discovery of this new universe confronts humanity with the reality that there is no centre. There are an enormous number of galaxies, in expansion. However, where is the centre? If we see these galaxies as belonging to the "world", we have to realise that there is no centre. The evolution of Europe as an idea of universalism could help us to surmount the fear of living without a centre.[75] We live on the surface; the centre of the globe has no conditions for living. The process of globalisation leads to the consciousness that we should not fix ourselves on one global centre of organisation and animation. To develop relations of sharing, maybe we should change the orientation of our perspective. Maybe the ever-growing knowledge of infinity leads us to a deeper inward knowledge that the centre does not exist outside, but inside man; interiority based on behaviour and sentiments of abundance, goodness, cordiality and benevolence.[76] Here we are somehow back at Plato's care for the soul: Peace is in man and between men, transgressing existing frontiers.

Søren Kierkegaard[77] re-thought this "original" idea. He distinguished several shifts in the development of human existence. He argues that the final phase in this process is the realisation of freedom. In his approach one starts with what he calls the "chaotically spiritless". This spiritless can be expressed via pagan oracles or sacrifices.

In a second phase, one becomes desperate about "earthly" aspects like business, age or talents. The individual traverses: one tries to "design" something like a "final sense of life". One tries to be "proud" of one's "self". This first traversal is characterised by rising doubts and felt as full of conflicts. An important shift can be made afterwards, a movement towards a new realisation of freedom. According to Kierkegaard, this shift starts with a kind of "instinct", a kind of primitive sense of immediateness. This becomes "framed" via education and insights. A new sense of infinity awakes, is first felt, then known and finally willed. In a last movement, man can discover a new type of existence, a revival of one's past and consciousness "in the presence of the spirit". This spirit leads to peace.

2.3.4 Space: universalism and the aerospace sector

"We also need an evolution towards a truly Europeanisation of science and technology, which, at this moment only exists in one segment: space".[78]

2.3.4.1 Universalism

We have seen that Europe as a figure evolves through a specific dynamics: universalism and withdrawal from the desire of territory. Europe became conscious of her "romanity", the idea that her identity is based on something "secondary". Europe arose out of what was "strange" to her. She desires to "transfer" things: objects, ideas, human beings etc. In this sense the problem of "identity" is a delicate one: on the one hand there is a kind of political and industrial will to define a European specificity and more particularly specific governance, on the other hand the most specific feature seems to be "transition". Europe tries to find a political and geographical form but at the same time, she wants to transgress and go beyond this concept of form: Europe wants to defend the idea of the "world as a world".[79] This particular dynamics of trying to grow from national to European "identity" and at the same time "giving up" a specific "identity" through the idea of becoming "the world", can also be recognised in the development of the aerospace sector.

However, Europe lacks a clear vision on projects in the space sector. Nevertheless, is there something more universal than space? When we speak about Europe and space, we qualify the universal aspect of space as "European". In this sense, "Europe" is a directive and regulative idea.[80] So, can the development of the aerospace sector in Europe be seen as a step towards universalism: presenting her mission and objectives as a project of peace and as a means or a medium to grow to a cosmopolitan possibility of world citizenship? Might "space"[81] become a new

"European frontier", a federative project for European institutions, the nation States, companies and citizens? Thanks to space faring, the Earth can be seen from the outside: this makes us conscious of the vulnerability of this planet. At the same time, only the human is capable of discovering this. Travelling into space offers us a new opportunity in our quest for humanisation. Does Europe dare to deepen her political will and vocation, not only serving her own interests but also daring to continue exploring space for the benefit of humanity by defending universal interest in the development of space projects?

2.3.4.2 The aerospace sector: cooperation and innovation

The Lisbon Treaty presents the European aerospace sector as one of the industrial segments where the ideals of cooperation and innovation can be realised. In Article 189, this Treaty specifically expresses the wish to develop cooperation: the political wish to grow to a stronger place in the worldwide setting is stated. The different European Councils[82] confirm regularly the same objectives: more competitiveness and innovation to become the strongest economy. However, this desire to promote innovation seems more a Grail for the 21st century than a well-framed working concept. Involved institutions are looking for a "European" legitimacy. At the same time, projects are already "universal" in the sense that partnerships are taking place worldwide. The aerospace sector is an example of this "Europeanness". It is a double transfer: an effort to become universal and a regular withdrawal from its concept of continentality. Europe as a figure expresses an attitude of "openness". She became conscious of the fragility of this attitude. She realises that the conquest of space can become a dangerous dream, not leading to innovation in the development of cooperation. Besides, the opportunity to use space activities to reflect upon the responsibility of the human in the Creation does not seem integrated in this process. The leading role of Europe could be in showing that ruptures form an essential part of new unities, new stabilities and more peace. How can the different partners in this process in the aerospace sector describe the pacification of this tension? Will they be capable of finding a specific European way to proceed to universalism?[83]

European policy should distinguish itself from other policies by showing the political willingness and offering the means to work in space by more clearly expressing the orientation of its space activities. Europe's fragility, namely not having an independent defence policy, can be used to implement the non-commercial and non- military uses of space and to create space projects as a good servant instead of a bad master of creation. Europe's role might be to open space for everybody.[84]

2.3.5 Conclusion

"With more humility being more ambitious"[85]

In this article, I have tried to investigate the question of Europe's "identity".

Europe is evolving as an economical and political entity, representing a specific concept of universalism in the world. The dynamics of her universalism is to become worldwide. Europe is more than a territory or a space. Europe is first an Idea of direction and regulation. Europe knows that she is a fragile "mi-lieu" without a centre. She transports ideas and goods and invites reflection upon the content of "the common good". Europe is an Idea: striving to Eternal peace in opposition to violence and destruction.

Europe is not just a geopolitical unity, expressing an advanced form of economic power in a polycentric world: Europe is a desire. She is conscious of the fact that she was born out of transport and ashes. Her "identity" came from outside herself. Europe is willing to incarnate that which comes from outside her borders. She searches for modalities to embody these ideas and to transport them further in the world. Europe is transport and knows that she is secondary. She is the tension between a denial of herself and the "going beyond" herself.

Ruptures and renaissances were needed to learn that an attitude of fear can be transformed into one of hope and revelation. Europe was born out of conquests and she strives for the Idea of eternal peace. Europe's principal role in the world could be to express and transport more competition to develop hope, peace, the joy of living and sharing.

The following suggestions could help achieving this modest but ambitious plan:

Reinforcing political and industrial will

European space policy should dare to express not only the profitable development of industrial activities in space. The space policy as well as the space industry should also express the nature of European identity, namely showing that it is looking for a universal project that develops peace. Means could be to create transitional spaces (workshops, symposia, trainings) in the different European countries to develop a common European industrial policy in space.

Education in space technology

Europe lacks an interest by students to study space engineering. Necessary is the development of curiosity by proposing projects, inspiring students to become engineers in this sector.[86] In education, this can be implemented by creating learning communities to develop consciousness of the role that space projects can play in the future. These projects should also elaborate ethical aspects: improving the state of nature, of communities and of individual man.

Peace and the humanisation of space

The progress of technology in space is not only a technological question. This development makes us conscious of the infinity of space. Space invites the human to reflect upon his responsibility to use the new technological opportunities. Europe's vocation is to offer peace to humankind by clarifying "the common good". This is what Europe defines as universalism. It reminds the human that he is a servant in the process of Creation.

[36] I would like to thank Prof. Dr. Donald Loose for his precious help and support to develop my ideas.

[37] Bidalon, Philippe. "Charles Martel, Un succès pas si franc." L'Express no. 3051–3052, Decmber 2009: 102–103.; Sénac, Philippe. Carolingiens et al-Andalus VIIIème-IXème siècles. Paris: Maisonneuve et Larose, 2003.

[38]Beatrix, Queen of the Netherlands, "Voorbeschouwing". Ornstein, L., Breemer, L. Europa. Grote denkers over Europa. Amsterdam: De Bezige Bij, 2007: 9.

[39]See the vigilance of H. de Regt for every form of essentialism: De Regt, Herman. "Europa bestaat niet." De Europese Europeaan. Schiks, Judith A. (ed.). Tilburg: CWL, 2005: 83–93. Tony Judt shows masterly how this trap leads to catastrophes: Judt, Tony. Postwar, A history of Europe since 1945. London: Vintage Books, 2010. Patočka, Jan. L'Europe après l'Europe. Paris: Editions Verdier, 2007.

[40]Guénoun, Denis. Hypothèses sur l'Europe. Belfort: Circé, 2000.

[41]Guénoun, Denis. "L'Europe et son idée." Eurostudia 1.2 (2005): 45.

[42]It is noteworthy to remark that etymologically the name Europè means: she who has the right or good (*eu*) sight (*opè*).

[43]Ovid. Metamorphoses, Book ii, 833–875, Book iii, 1. Trans. A.D. Melville. Oxford: Oxford University Press 1986. Also: Horatius. "Verzamelde gedichten". Groningen, Historische Uitgeverij 2003: 378–383. For a recent comment, see: Berns, Gido, "Wij, Europeanen", Best: Damon 2005: 7–17. However, elsewhere the "ancestor" also came from the "outside": Abraham, Quetzalcoatl, Aeneas.

[44]Guénoun, Denis. Hypothèses sur l'Europe. Belfort: Circé, 2000: 35. Except Herodotos, the word *Europe* occurs rarely in Greek texts: only some nine times. *Ibid.*: 33, note 20.

[45]*Ibid.*: 49–50.

[46]*Ibid.*: 44.

[47]Patočka, Jan. L'Europe après l'Europe. Paris: Editions Verdier 2007.

[48]Guénoun, Denis. Hypothèses sur l'Europe. Belfort: Circé 2000: 60.

[49]*Ibid.*: 54. ; Brague, Rémi. Europe, La voie romaine. Paris: Gallimard 2005: Chapter IX.

[50]Guénoun, Denis. Hypothèses sur l'Europe. Belfort: Circé 2000: 74–75.

[51]*Ibid.*: 66–67.

[52]On frontiers and borders: see Abel, Olivier. La justification de l'Éurope. Essai d'éthique européenne. Génève: Labor et Fides 1992: 67–85.

[53]Guénoun, Denis. "L'Europe et son idée." Eurostudia 1.2, 2005: 97.

[54]"Space" here is meant metaphorically: the space of mind.

[55]Loose, Donald. „Rien du tout. L' Espace européen." Paper for the 2nd Round Table Conference „Jean Monnet", Paris, Observatory 2008.

[56]„Space" is used here in the classical astronautical sense.

[57]It is worth noting that the leading music of this film, the expression of the movement of satellites, is the Blue Danube, composed by Johan Strauss.

[58]Clarke, Arthur C. 2001: A Space Odyssey. Arthur C. Clarke and Polaris Publications, Inc., 1968.
[59]Kant, Immanuel. „Zum ewigen Frieden. Ein philosophischer Entwurf." Berlin Leipzig, Akademie Ausgabe, VIII: 341–386.
[60]"Space" here is meant in the geographical sense.
[61]Barnavi, Eli. "Mille ans de construction européenne." Barnavi, Eli and Paul Goossens (Eds.). Les frontières de l'Europe. Brussels: De Boeck&Larcier, 2001: 29–30.; Luypen, W.A.M.: "The West has made three major decisions: first, knowledge is scientific knowledge; second: the state is a constitutional state; third: life is a life of labour." Lectures Spring season 1974, Tilburg University.
[62]The norm of peaceful uses of outer space is also embedded in the United Nations space treaties that were drafted in the 1960's and 1970's (i.e., before the creation of the ESA). In addition, virtually all the space faring nations highlight their peaceful ambitions in their official space-related documents. However, the validity and singularity of Europe as a geopolitical unity striving for peace is the fact that Europe is the only supranational entity able to define this purpose.
[63]However, only 3% of the European population works in country other than its own. It is only a very small group that circulates: this is what we can call the « Europe-club ».
[64]Wihtol de Wenden, Catherine. "Migrations, pluralitè, intègration." Paper for the 1st Round Table Conference "Jean Monnet", Houjarray, 2007. Europe makes a distinction between territory and space: territory can be traversed; European space is trans-national. Territory is connected to state formation, governments and political population. In this sense Europe transgresses the use of space to define a national identity. Space allows citizens to evolve. Democracy is the incarnation of the will of citizens. These distinctions are evolving. This is why it is so difficult to define European "identity".
[65]"Space" is used here in the geographical, territorial sense.
[66]Some discussion is possible about the democratic legitimacy of the European Court for Human Rights: cfr. Baudet, Thierry. „De gespannen verhouding tussen mensenrechten en rechtsstaat". Harchaoui, S. and Jonkers. J. (eds.), Leve de rechtsstaat. Amsterdam: Boom Lemma, 2010.
[67]However, the decision mechanism of the EU is still a mixed polity, partly supranational and partly intergovernmental.
[68]Patočka, Jan. L'Europe après l'Europe. Paris: Editions Verdier 2007: 34.
[69]*Ibid.*: 28–36.
[70]Husserl, Edmund. Die Krisis der europäischen Wissenschaften und die transzendentale Phänomenologie: Eine Einleitung in die phänomenologische Philosophie (1936). W. Biemel (ed.), Husserliana, Band VI. Den Haag: Martinus Nijhoff 1976.
[71]*Ibid.*: 28.
[72]*Ibid.*: 29–30.
[73]*Ibid.*: 31.
[74]Doré, Joseph. "Au cœur du monde." La mission: tous envoyés. Christus, 156, October 1992: 429.
[75]When there is no center, neither there is a periphery.
[76]*Ibid.*: 429
[77]Kierkegaard, Søren. Crainte et tremblement. Œuvres Complètes IV. Paris: Ed. de l'Orante 1972; Scholtens, Wim R. Kierkegaard als psycholoog. Baarn: Ten Have, 1979.
[78]Dupas, Alain and Huber, Gérard. La grande rupture? L'humanité face à son future technologique. Paris: Robert Laffont 2010.
[79]Guénoun, Denis. Hypothèses sur l'Europe. Belfort: Circé, 2000: 293.
[80]Loose, Donald. "Rien du tout. L'Espace européen." Paper for the 2nd Round Table Conference "Jean Monnet". Paris, Observatory 2008: 2.
[81]"Space" is used here in the classical astronautically sense.
[82]The Cologne European Council (1999) recognised the need for sustained efforts to have a competitive and dynamic industrial and technological defence industry in support of Europe's capacity to respond to international crises. At the Lisbon Council (2000), the Heads of State and Government set the Union the strategic goal of becoming the most competitive and dynamic knowledge-based economy in the world within the next decade. This message was reinforced at the Barcelona Council (2002), which called for a significant boost in the overall R&D and innovation effort in the Union.

More recently still, the Thessaloniki Council (2003) decided that the time had come to take concrete steps in the defence field. A globally competitive aerospace industry is central to the achievement of Europe's economic and political objectives.

[83]See, for example the Criteria of Copenhagen (1993) where several aspects of a "European" identity are expressed. There are no geographical limits, nor linguistic, religious or racial criteria to define the identity of Europe. Europe is an effort to find a form of universalism, a process of expansion through an extension towards new physical spaces and via a transfer towards new universes of meaning.

[84]See the intervention of Hyppolite Simon, 2nd. Round Table Conference "Jean Monnet". Paris, Observatory 2008.

[85]Moïsi, Dominique. „De emotionele grenzen van Europa". Ornstein, L., Breemer, L. Paleis Europa. Grote denkers over Europa. Amsterdam: De Bezige Bij 2007: 122.

[86]Cfr. the intervention of Johannes von Thadden. 2nd. Round Table Conference "Jean Monnet". Paris, Observatory 2008.

Space and the society

2.4 European identity through space – how to make public opinion instrumental[87]

Marcus Hornung

Space could significantly contribute to the creation of a European identity. This is obvious screening the articles in this volume and was demonstrated in the course of the "European Identity Through Space" conference held at ESPI on 12–13 November 2009, when several case studies were presented on this subject. Space, with its numerous applications, has become an integral part of everyday life, including arts, exploration, security, defence and the economy. This article aims at extending this debate on European identity through space to the sphere of public opinion. Space will only make its way into the formation of European identity if the European people are able to identify themselves with space applications, increase loyalty towards space research and at the same time profit on a personal level from the benefits of its use. The article therefore focuses on two main points: firstly, it recalls once again the importance of the space sector for a better understanding of the European identity and, secondly, it introduces a so-called participative component of European identity.

"There are so many different approaches to the concept of [. . .] identity [. . .] in the EU that it has to be called into question if all authors mean the same thing. A credible answer would be: No."[88] concedes the Italian political scientist Furio Cerutti. A well-founded discussion on European identity requires clarifying the terms in use.[89] Two underlying conceptions of European identity can be identified within the myriad of ideas on how to operationalise it: first, a definition of European identity based on common characteristics such as a retrospective vision;[90] second, an understanding of European identity as self-perception by European citizens – the participative dimension.

The projective approach assumes that European identity can be captured by a simple description of Europe's character. It includes a pre-supposed common heritage of a historical, political, technological, philosophical and cultural nature.[91] The common European history, or the set of values based on democracy,[92] are only two of many examples of identity shaping elements. However, some points of criticism of this retrospective understanding have to be mentioned:[93] the concept is extremely flexible, subject to ongoing changes and different political points of view, and therefore hard to grasp for researchers. Hence, researching European identity involves more than merely discovering common cultural substance heritage[94] and applying it to concrete fields of interest, such as a European Space Policy.

The second approach towards European identity focuses on the relationship between European citizens and the EU itself within its various fields of action and politics – a European identity formed by the self-reflection of Europeans themselves.[95] This collective identity is created through a cognitively shaped connection to a certain objective. Additional to this affective dimension of emotional connection, Nissen[96] supplements the debate with the introduction of a utilitarian component: for a sustainable identification with Europe, simple loyalty and empathy as well as the material and user-oriented dimensions are necessary. This implies an enduring connection between the interests of European citizens and the EU itself. This could be a case of Europeans profiting on a personal basis from policies at the European level. European people's views, opinions and perceptions are the integral elements of European identity. Therefore, public opinion has to be taken into account when applying the concept of European identity to the example of the European Space Policy.

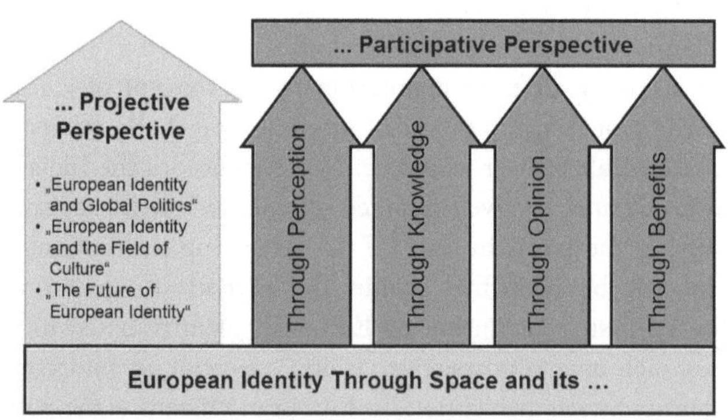

Fig. 2.5. *European identity through space and its projective and participative perspectives.*

2.4.1 European identity through space and its projective perspective

The 2009 conference on "European Identity Through Space"[97] proposed in its programme to cover "technical" (technology, applications, security, industry, economy) as well as "cultural" (literature, film, arts, education, philosophy) fields of research. In doing so, its approach to European identity corresponded more to the projective one: the participants investigated the effect space has already had on shaping a common European identity. Derived from these analyses, the speakers evaluated how space can further contribute to form a stronger European identity. This interdisciplinary approach allowed in particular overcoming the limits previously set regarding this debate. The conference outcome could be categorised in three distinct areas, or "spaces":

"European Identity and Global Politics" offered suggestions on how space technologies and applications can increase public awareness of European identity. The speakers argued for example, that European space industry creates space technologies like the successful Ariane launcher. Such flagship programmes function as the symbols of the scientific, industrial and technological leadership of Europe in the world. Such leadership creates a clear distinction between European and non-European space actors, which in turn is a driving force behind forging a sense of common European identity in itself.[98]

"European Identity and the Field of Culture" reviewed the influence of space on the humanities within numerous disciplines. Using examples from art, from the figurine "Cosmic Dancer" to "Star Wars", enabled overcoming classical boundaries to stimulate creative thinking. Culture in general is an enormous identity-shaping element on a European level. The fact that space found its way into European culture within the past centuries was proven by the contributions to the conference.[99]

Finally, "the Future of European Identity" summarises the possible future effects of European space accomplishments on European identity. This third "space" proposes to extend the dialogue on space issues and reflect upon space achievements more intensively in order to further strengthen the influence of space on the formulation of a European identity.[100]

In a nutshell, the greatest advantage of the 2009 conference lay in applying the projective approach to forging a European identity through space. The final conclusion, which was that European space activities can indeed be seen as a step towards European identity, was itself a valuable achievement. However, most conference contributions used either a descriptive or retrospective line of arguments. Of all the participants, only Etelka Barsi-Pataky extensively touched upon the issue of public opinion during her examination of what Europeans might expect from space.

2.4.2 European identity through space and its participative perspective

The participative approach includes on the one hand the affective and on the other hand the material and utilitarian dimensions. The constructive influence of space on European identity would be vindicated if European space activities prove convincing to European citizens in the cognitive and emotional sphere as well as in the utilitarian sphere.

The only way to verify this statement is through the help of data derived from public opinion polls. The most comprehensive up-to-date data collection is provided by Eurobarometer in its 2009 flash report "Space Activities of the European Union".[101] For findings to be conclusive, Europeans would have to identify themselves with European space activities. Depending on their respective level of cognitive and emotional bond, opinion poll answers can be categorised according to whether they imply a simple perception, a definite knowledge, or even a clear opinion on European space activities.

2.4.2.1 Perception

Following this approach, it has first to be established whether Europeans really perceive the existence of European space activities or not, as this would constitute a minimal level of cognitive attention. To approach this dimension, Eurobarometer uses the question "Have you heard about the European Earth Observation satellites?", which suggests Earth observation as a sample and does not include any value judgment. Within the EU-27, 56% responded that they were informed about the existence of such space applications on a European level. 42% however had never heard of European Earth observation capabilities. Furthermore, it might be interesting to note that the degree of awareness towards European space application varies between EU Member States: whilst 76% of Slovenes are aware of the system, only 26% of Swedes know about its existence.[102]

How can one evaluate these results? Earth observation is not just a random European space application project, but one of the flagship programmes under the European Space Policy, namely GMES. It poses an adequate case study example, because Galileo, in contrast, is a broadly discussed and well-known achievement and would therefore not deliver significant results for public opinion research purposes. The lion's share of the space research spending within the 7th EU Framework Programme (which constitutes only one out of many sources of public spending in this sector) is dedicated to Earth observation.[103] Finally, Earth observation contributes to numerous operational areas in our everyday lives. Taking this into account, it is extremely surprising that nearly half of all Europeans are not aware of the existence of European Earth observation capabilities. This awareness gap constitutes an insufficient basis for both knowledge of and positive attitude towards European space activities. The enormous discrepancy between the political and economic significance of European Earth observation capabilities and the level of public awareness surrounding them does not yet allow us to consider space as an identity-shaping element in this field.

2.4.2.2 Knowledge

As a contribution to the formulation of a European identity through space, it is even more relevant to take into account European public knowledge of European space projects: acquired knowledge of space activities is certainly the second requirement for an affective bond between the European public and space activities. As far as Earth observation is concerned, only 61% of Europeans have a positive knowledge of earth observations systems, or even know "what they do".[104] No more than 34% of the people asked knew what the term "Earth observation" means. The probably better known satellite navigation system Galileo, which was also covered by Eurobarometer surveys in 2007, gave some more positive results. When asked if they knew "what is satellite-based navigation?",[105] at least 68% knew what these systems were used for. Knowledge about European space applications programmes is usually more developed among men than women. Surprisingly, knowledge about European space projects is least developed among young Europeans between the ages of 15 and 24.[106]

With the help of these samples, the following conclusions can be drawn: first, there are significant differences in terms of knowledge of space application programmes, depending on their type. This can be explained for example by the fact that Europeans benefit from satellite navigation services directly in their everyday lives, something which is not so much the case for Earth observation applications. The considerably high ignorance of the younger age group leads to the conclusion that space education is not included in the classic primary and secondary education curricula, but is instead usually of an extracurricular nature (e.g. out of personal interest and experiences, or through the media). Taking all these points into account, the level of space knowledge does not really allow us to establish a link between space activities and European identity.

2.4.2.3 Opinion

The dimension of knowledge on space activities did not include an evaluation of its emotional impact. Consequently, the next step was to come up with indicators in order to assess the emotional attachment of Europeans towards the European Space Policy. More precisely, estimations were made as to how important various components of European space activities are for the European citizens. When asked "How important is it in your view to develop the following space applications for Europe?",[107] 90% of Europeans chose Earth observation, 75% communication applications and 67% satellite navigation as the most or more important to them. Furthermore, these surprisingly positive results are supported by 74% of persons interviewed, who stated that Europe should even increase its international engagement in space policy matters.[108] However, this positive stance was somehow limited by the usually high costs of space research and services: only 20% of Europeans would support an extension of the share of space policy spending within the EU budget, in accordance with the objectives of a more ambitious European Space Policy.[109]

In order to properly evaluate these results, it has to be recognised that the fact that integrating the generally expensive space applications systems on a European level ultimately saves public funds (compared to the individual Member State's projects and as long as they are embedded in a functioning cooperation framework) does not seem to be widely understood. In this respect, the quantitative results of the survey ignore the qualitative advantages of a common European Space Policy. Additionally, it is probable that Europeans would refrain from increased spending in areas that are not perceived as essential, such as research and development, culture or outer space because of general public spending shortages caused by the economic and financial crisis. However, when taking the other research results into account it can be concluded that European public opinion towards European space projects is a very positive one. Most of the people interviewed stated for example that Earth observation, communication and satellite navigation systems were indeed important for them and that they would even like to see the EU strengthen its space policy efforts. Therefore, one can conclude that European public opinion has a rather constructive and supporting attitude. Together with the perception and knowledge dimensions, this could be a small step towards a significant contribution of space to the European identity. It seems that a cognitive-emotional bond between Europeans and European space projects does in fact exist. However, it was also demonstrated that knowledge of space activities is very modest, although there is a generally positive public attitude towards European space efforts.

2.4.2.4 Citizen's benefits

In his introductory statement to the conference "European Identity Through Space", ESA Director General Jean-Jacques Dordain stated: "I have a dream of a 'day of outer space'. On this day, all satellites should be switched off, so that people recognise how important space technology is. No Parisian taxi driver could survive without outer space".[110]

In his speech, he explained that there is a necessity for a cognitive-emotional as well as a utilitarian bond to space activities, in order for space to make a significant cohesion-building contribution to European identity. This could be achieved if the individual citizen recognises that space activities create benefits for him on a personal level. This is why ESA Director General Jean-Jacques Dordain emphasised the relevance of space applications for each and every European, even if this fact is not always obvious to Europeans themselves.

Through its 2007 study entitled "Case for Space – Space Applications meeting Societal Needs",[111] ESPI has provided several examples of areas in which space not only affects European society as a whole, but also the common man on the street. Eumetsat, for example, guarantees the everyday availability of precise meteorological forecasts for European citizens – a highly complex service, which is not scrutinised. Geographic Information System (GIS) permits satellite-based traffic management and orientation in major urban centres, which are indispensable but not commonly known. Eutelsat enables high-speed internet access in rural areas that are not connected to ordinary broadband land lines. Television via satellite is also taken for granted. Many more examples could be given. In any case, there is no doubt about the fact that space technologies affect the everyday life of each and every European.

While this fact is already clear to space policy actors and analysts, it has not yet been fully appreciated by European society in general. Awareness of the close linkages between space activities and the daily routine of citizens is one of the major requirements on the way towards a European identity through space. However, what would happen if space benefits were made available without Europeans noticing them, as it is the case for most of the services that we use and that we take for granted? The result would be a positive but rather reluctant reflection of European space activities in public opinion. Especially against the background of the aforementioned public opinion polls, an imbalance between European space awareness and space dependence has to be taken into account.

2.4.3 Conclusions and recommendations

Space already contributes to European identity in its projective perspective. In order to constitute an element of the European identity in its participative approach, it is indispensable to close the gap between the public's low awareness of space activities on the one hand and its great dependence on them on the other. This article has demonstrated so far that a cognitive-emotional bond between Europeans and European space efforts does exist, but that it is nevertheless poorly developed. Even the user-oriented relevance of space applications within the context of Europeans' daily lives cannot diminish this problem. This leads to several recommendations.

2.4.3.1 Improve public perception of space benefits

A main finding of this article is that Europeans are not aware of the relevance of space applications to their everyday lives: the built-in GPS systems guiding countless cars worldwide and the usage of free Earth observation data via internet are only two of the dozens of highly complex applications that are easily taken for granted. Convincingly communicating the real role of space in modern society may be challenging, but it is nevertheless essential.

Indeed, European Space Policy actors such as ESA have already recognised the need for action and they have taken steps to improve public perception of the benefits of space activities: for example, "The Impact of Space Activities upon Society"[112] was a successful project conducted by the International Academy of Astronautics (IAA) in cooperation with ESA. The corresponding publication[113] is a great example of giving the public an idea of how they can make use of space applications: more than 70 interdisciplinary experts with different space-related backgrounds completed the sentence "I believe that space activities are impacting society through . . . " from their personal point of view. Additionally, ESA installed a website category named "Benefits for Europe",[114] explicitly explaining the benefits of space exploration to the interested public. Another example of how to bridge the gap is given by the Paris based NGO Eurisy,[115] whose members are space agencies and governmental or intergovernmental organisations. Its mission is to enable society's access to satellite information and service benefits by raising awareness and providing support, advice and information.

Undoubtedly, these initiatives should be regarded as ambitious case study examples. However, such rudimentary efforts are the starting points for further engagement with public opinions. A major improvement would be to make publications such as "The Impact of Space Activities upon Society" available in all EU/ESA Member States' languages. The approach of relying on English as lingua franca will ultimately fail, if the aim is to improve the common man's awareness of space activities' benefits. Even complex image campaigns would be effective ways of communicating on the sustainable improvement of public perception of space benefits. Other high technology sectors such as chemistry or pharmacy could set the example for these efforts. In the field of space policy and industry, public relations work has to expand from potential end-users to the entire society. Without trying harder to improve the public perception of space benefits, space won't become part of the European identity in its participative dimension. Finally, successful discussion fora such as this year's International Space University symposium on "The Public Face of Space"[116] could help to identify new ways to improve the public perception of space benefits.

2.4.3.2 Improve space education

The starting point for creating more positive attitudes through increased educational activities on space lies in the necessity of creating knowledge about, in addition to perception and support for, this specific element of European identity. Public opinion poll results have shown that there is a serious lack of knowledge concerning European space projects, applications and policy. The public's knowledge on space is rather modest, especially among younger people. One reason for this is that some EU Member States' school systems seem to pay only little attention to subjects such as astronomy. Additionally, there are huge differences between the different European Member States concerning knowledge of space applications. As education is traditionally an area of national competency, there is no way to initiate legally binding steps on a European level.

However, it is up to the European Commission to appeal to the individual Member States in order to improve space education within each and every national school system. The 2008 European Commission communication "Improving Competences for the 21st Century: An Agenda for European Cooperation on Schools" explicitly takes into account the need to increase understanding of modern science and technology.[117] A next step could be to identify space research as a priority area within these efforts. Why not strongly implement space research as an obvious case example for applied sciences in our pupils' natural sciences curricula? The main focus of such ambitions should indeed be on general space education in our public schools: extracurricular projects already exist for pupils and students, showing explicit interest in space related issues.[118] But only if space education in general is improved, can a sustainable European identity in its participative perspective be taken seriously.

2.4.3.3 Improve the attractiveness of space

Once perception and knowledge of the relevance of space projects to everyday life have been raised to a satisfactory level, there would still be one building block missing in order to achieve a strong cognitive-emotional bond. No matter how much information and education are disseminated, this will not lead to every European being automatically infected by a kind of "space fever". A positive attitude and opinion on space projects and even towards related public spending has to be ensured.

As a matter of course, each European has to make up his own mind regarding his stance on European space projects. Nevertheless, it is possible to promote space to the public by disseminating the experience of space research and exploration. Both public and private space actors could make good use of the fascination surrounding outer space. One practical proposal is given by Harris[119]: citizens should be able to get an impression of how space applications work in reality, e.g. by field trips to facilities such as launch sites, ground control segments or visits to theme parks related to space. According to the author, such visitor centres already exist in the U.S. and China. Why not allow such guided tours into non-security relevant facilities in Europe as well? Another way to promote space fascination lies in the development of commercial space tourism. Although space tourism is aimed at a limited number of people who can afford it, it nevertheless enables all citizens to emotionally relate to space exploration, which helps to bridge the aforementioned imbalance. Creating a sustainable positive attitude towards European space projects would be the final requirement for building the cognitive-emotional bond needed for the participative element of European identity.

2.4.3.4 Improve transparency of European space projects

As public opinion polls reveal, there is a lack of willingness to invest public funds into space activities on a European level. The main reason for this is the absence of general understanding of the pros and cons of an integrated European Space

Policy jointly run by the EU and ESA. In this respect too, public opinion can be instrumental. Improving transparency regarding public spending on European space projects is an instrument for increasing information and education on space, and shaping public attitudes towards space activities at the same time. It is up to the EU and ESA as the principal actors of the European Space Policy to explain some of the aforementioned facts to Europeans via the channels of political communication.

These would include understanding that space projects are indispensable to our every day lives and this is one of the reasons why they are so expensive; that well integrated space capabilities and applications on a European level would nevertheless save money compared to conducting operations on a national level; that in order to make the step towards partial European integration possible, a certain budget would be needed. It is first and foremost the task of the European Commission and ESA to promote the European Space Policy in a transparent and easily understandable fashion, open to each and every European.

Up to now however, it seems as if the European Commission underestimates the identity shaping power of European space projects. Scanning the set of documents leading up to a coherent European Space Policy, one can barely detect references to the development of European identity through space. One reason is for instance that even policy makers such as European Commission's President Barroso refrain from speaking of the symbolic value of space and seem to take everyday space benefits for granted: "[...] [S]ometimes politically we speak about space, people believe that we are speaking about some priority very distant in the future. And I think it is very important [...] to underline that when we are speaking about space policy, we are not simply talking about a symbolic, a great project for the future, it is about very concrete applications for the life of our citizens [...]".[120] But what if our citizens are not even aware of space's benefits?

To conclude, there is still need for action. The above-mentioned steps could help raise the acceptance of the European Space Policy significantly, as well as help to achieve the sustainable development of a European identity through space, not only in its projective but also in its participative point of view.

[87] This article is an updated version of an ESPI Perspectives from June 2010: Hornung, Marcus. "European Identity through Space – How to Make Public Opinion Instrumental" ESPI Perspectives 36. Vienna: European Space Policy Institute, 2010.

[88] Cerutti, Furio. „Warum sind in der Europäischen Union politische Identität und Legitimität wichtig?" Europäische Identität als Projekt, Innen- und Außensichten. Eds. Thomas Meyer, Johanna Eisenberg. Wiesbaden: VS Verlag für Sozialwissenschaften, 2009: 250.

[89] Schmale, Wolfgang. „Geschichte und Zukunft der Europäischen Identität“ Bonn: Bundeszentrale für politische Bildung, 2010: 38. Also see the contribution of Mathias Spude in this volume.
[90] Thiesse, Anne-Marie. „Die europäische Identität: Erbe der Vergangenheit oder Konstruktion für die Zukunft?“ Europäische Identität als Projekt, Innen- und Außensichten. Thomas Meyer, Johanna Eisenberg (eds.). Wiesbaden: VS Verlag für Sozialwissenschaften, 2009: 32.
[91] Schmale, Wolfgang. „Geschichte und Zukunft der Europäischen Identität“ Bonn: Bundeszentrale für politische Bildung 2010: 31–32.
[92] European Commission. Public opinion in the European Union. Standard Eurobarometer 71. 2009. 21 November 2010. *http://ec.europa.eu/public_opinion/archives/eb/eb71/eb71_std_part1.pdf.* 39.
[93] Dauderstädt, Michael. „Der Kirchtum und sein Horizont, Identität und Grenzen Europas.“ Bonn: Eurokolleg der Friedrich-Ebert-Stiftung, 1999.
[94] Loth, Winfried. “European Identity in a Historical Perspective” ZEI Discussion Paper. Bonn: Center for European Integration Studies, 2002. 21 November 2010. *http://www.zei.de/download/zei_dp/dp_c113_loth.pdf.* 1.
[95] Meyer, Thomas. “Europäische Identität“ Europäische Identität als Projekt, Innen- und Außensichten. Thomas Meyer, Johanna Eisenberg (eds.). Wiesbaden: VS Verlag für Sozialwissenschaften 2009: 15.
[96] Kraft, Ekkehard. „Lang lebt das Vaterland, Die europäische Identität – eine Chimäre?“ Neue Züricher Zeitung, 14 July 2002. 21 November 2010. *http://www.nzz.ch/2002/06/14/fe/article86ETC.html.*
[97] Nissen, Sylke. „Europäische Identität und die Zukunft Europas“. Aus Politik und Zeitgeschichte 38 (2004): 21–22, 26.
[98] “European Identity Through Space – Keywords for the Future” 2010. European Space Policy Institute. 21 November 2010. *http://www.espi.or.at/index.php?option=com_content&view=article&id=456:2-february-2010-flyer-on-european-identity-through-space-online&catid=39:news-archive&Itemid=37.*
[99] See the contribution of Alan Belward in this volume.
[100] See the contribution of Giulia Pastorella in this volume.
[101] See the contribution of Jacques Blamont in this volume.
[102] European Commission. Space Activities of the European Union. Flash Eurobarometer 272. 2009. 21 November 2010. *http://ec.europa.eu/public_opinion/flash/fl_272-en.pdf.*
[103] *Ibid.*: 9.
[104] European Commission. Work Programme 2008, Cooperation, Theme 9, Space. COM(2007)5765 of 29 November 2007. Brussels: European Union: 41.
[105] European Commission. Space Activities of the European Union. Flash Eurobarometer 272. 2009. 21 November 2010. *http://ec.europa.eu/public_opinion/flash/fl_272_en.pdf.* 29.;
European Commission. General public survey on the European Galileo Programme. Flash Eurobarometer 211. 2007. 21 November 2010. *http://ec.europa.eu/public_opinion/flash/fl_211_en.pdf.* 6.
[106] *Ibid.*: 28.
European Commission. Space Activities of the European Union. Flash Eurobarometer 272. 2009. 21 November 2010. *http://ec.europa.eu/public_opinion/flash/fl_272_en.pdf.* 29.
[107] *Ibid.*: 10.
[108] *Ibid.*: 17.
[109] *Ibid.*: 18–19.
[110] “Total abstrakte Gebilde.” 18 November 2009. Der Standard. 21 November 2010. *http://derstandard.at/1256745057136/Total-abstrakte-Gebilde.*
[111] Doldirina, Catherine. “Case for Space – Space applications meeting societal needs.” ESPI Report 7. Vienna: European Space Policy Institute, 2007.
[112] ESA/IAA. The Impact of Space Activities upon Society Website. 21 November 2010. *http://www.spaceandsociety.org/.*
[113] “The Impact of Space Activities upon Society“ The International Academy of Astronautics and the European Space Agency. 2005. 21 November 2010. *http://www.esa.int/esapub/br/br237/br237.pdf.*

[114] ESA website. 21 November 2010. *http://www.esa.int/esaCP/Benefits.html.*
[115] Eurisy website. 21 November 2010. *http://www.eurisy.org.*
[116] "The Public Face of Space". 14th ISU International Symposium. 16–18 February 2010. Strasbourg, France. 21 November 2010. *http://www.isunet.edu/index2.php?option=com_docman&task=doc_view&gid=1036&Itemid=26.*
[117] European Commission. Communication from the Commission to the European Parliament, the Council, the European Economic and Social Committee and the Committee of the Regions, Improving competences for the 21st Century: An Agenda for European Cooperation on Schools. COM(2008)425 of 3 June 2008. Brussels: European Union: 3, 6.
[118] E.g. the Space Education Institute Germany (*www.spaceeducation.de*) for pupils of all ages or the well known International Space University France (*www.isunet.edu*) for university students
[119] Harris, Philip R. "Space theme parks: Promoting space to the public" Space Policy 25 (2009): 88–89.
[120] Barroso, José Manuel. "The Ambitions of Europe in Space." Speech. 15 Oct. 2009. 21 November 2010. *http://ec.europa.eu/enterprise/policies/space/files/policy/the_ambitions_of_europe_in_space_en.pdf.*

Opinion

2.5 European identity through space

Résumé of the conference

Jacques Blamont

According to the introduction of Dr K.U. Schrogl, the subject of the conference was to consider the question: how can space become a part of European identity? Or, how can space contribute to the building of Europe?

This résumé will therefore follow that line and pick up the various suggestions made to try to answer the question.

Rather than summarise each talk one after the other as they were presented during the conference, I have chosen to incorporate the main ideas that surfaced from the speeches and discussions into my personal vision of European space.

2.5.1 Definitions

A number of presentations were devoted to delineating the specifics of the European identity. The following ideas were mentioned:

- Europe is made of a number of states. Its identity is a specific relationship between differences that become stabilised. It is a process, a passage, a mi-lieu.
- Different definitions: geographical, historical, political (EU, ESA, Council of Europe), conceptual (tradition since Antiquity).
- Major concepts: rationality as a paradigm and unifying heritage, citizenship law, enlightenment, culture, justice, trade, the idea of universalism.
- Culture of freedom.
- Culture of innovation, applications of science.
- Transgressing frontiers. There is a European concept of peace, conceived not only as the absence of war, but as a condition for opening out for men, a desire to move from fear to hope.
- Long history of partnership inside and outside Europe, among Member States and with the whole world.

- Europe is the biggest aid distributor (52% of total world development assistance), the largest trading block, the signatory to numerous multilateral environmental agreements, it carries out conflict prevention and crisis management missions throughout the world.
- Major problem: identity. *We have Europe, now we need Europeans (B. Geremiek).*

So, what is the connection between identity and space?

- Europe must have robust and effective space capabilities because of their importance to the economy, defence and communications. Without a space policy, Europe could become irrelevant.
- But this necessity is not recognised. Only 20% of Europeans consider space activity as very important, 40% important (the "very important" is a criterion for a real contribution to identity).
- Space in Europe has to have a unified voice.
- Europe has to create projects in space as a good servant instead of a bad master, and these projects have to become known and cherished by the public.

2.5.2 Concrete suggestions

When space started in history it performed a function that can be described as the sceptre of the Prince, a symbol of the power of the State, as I wrote years ago, but it has changed and become useful by generating many applications and services. It was the sceptre of the Prince in the Soviet Union and in the United States when Sputnik, Gagarin and Apollo created a strong sense of identity, and it plays the same role today in China and India, but not in Europe, because Europe is characterised by the absence (and rejection) of a Prince. One exception (which proves my point) was France under De Gaulle who, being a Prince, created the fist space programme in Europe, the Diamant launcher.

This absence explains why space in Europe has always been considered as a minor activity, with a budget today that is one sixth of the expenditure of the U.S.

The Director general of ESA, J. J. Dordain, told us of the importance of a new factor, the attribution of a space competence to the European Union by the now adopted Lisbon Treaty. This competence has to be used to promote space as an identity factor for Europe. We have to work on it!

The first priority is therefore *an overhaul of the European space institutions.* Created in the 1960–75 period, at a time when there was no European Union, no euro, no Astrium, no Ariane, no microprocessor, no mobile phones and no Internet, they are now obsolete and indeed are threatened by a movement towards the re-nationalisation of space, which has many examples including the lunar missions considered by Germany and the UK or even worse, in the rumoured joint German-French climatology mission.

A UNESCO ambassador once told me that when you ask an African: what is space? he will answer: NASA. What is space for European citizens? NASA. Space for Europeans has to be incarnated in their own visible Agency that would indeed be the sceptre of a Prince, the European Union. Institutional reform has to change the relationship between ESA and the EU government system (Council, Commission and the Armament Agency) with the absorption of the national Agencies into a rejuvenated ESA.

Such a change, which could be realistically undertaken in the post-Lisbon era, is an essential condition for building the image of space in the public and in the political sphere, for defining medium and long term strategies, for expanding R&D, for entering into new fields such as security and defence, for obtaining reasonable budgets and even for the implementation of projects.

Europe has to appear *focused on major programmes and projects*, clearly identified by the public. Let us quote some of the most important ones:

Fighting climate change is already seen as a European specialty, at least in the political arena, and materialises in a number of European space projects. However 73% or UK citizens and 72% of Swedish citizens for example, have never heard of European remote sensing programmes such as ERS or SPOT. It is not sure that any European taxpayer is aware of the SENTINEL programme.

A neighbouring domain is *aid to developing nations* (Europe, the largest donor, contributes 52% of total aid). Space should be used for promoting the European values of eradicating poverty, avoiding conflict and managing resources; a number of programmes exist already which could be oriented in this direction; for instance, remote sensing satellites would help passing to Africa our experience in agricultural monitoring.

In this context, GMES has to be redefined. It has been called an animal without a face. It does not include climatology; its security part is very obscure. Its policy towards customers from developing countries is confidential. Contents, objectives, frontiers are certainly not adequate to its status as a flagship of the European Space Policy.

No doubt that programmes in these areas could contribute strongly to European identity, but they don't. What is needed is a vision that would transform disjointed efforts into a clear strategy.

Security is an identity-forming element at the national level, but not at the European level. Space could help in the evolution towards a change of mind, since space has a security dimension, as security has a space dimension.

Europe could establish itself as a normative power able to generate codes of conduct.

Serious consideration should be given to the use of space for defence. Since the founding fathers of the pre-ESA space agencies ESRO and ELDO banished defence from the space activities in Europe, the art of war has been revolutionised, particularly in the 1990–2000 period, by space. Should Europe ignore the facts or adapt? It is certain that defence is also a strong identity forming element.

Towards the public: In the last five years, space applications have been adopted by unconscious millions.

Localisation: GPS has become the tool of hundreds of millions people. The other flagship of the space effort of the EU, Galileo, was born because localisation by satellite is too important to stay outside of any European control. But this strategic (and identity forming) character has been denied all along in the acceptance process, together with national manoeuvres leading to disastrous management. The project has to be reoriented with emphasis given again to its strategic nature.

Remote sensing: The production of images by satellites has been recently popularised by Google Earth. It is a sorry paradox that profits generated by satellites paid by European taxpayers are reaped by American portals. Here again, space could have an identity forming value provided it is properly exploited and promoted. Maybe the European failure originates in the technocratic grip over space. Management by public oriented leaders replacing engineers may be needed. Here we return to the GMES pending disaster.

Telecommunications is now essentially a commercial activity. Should the major European role (Arianespace launches more than half the telecommunication satellites and Astrium is one of the world leaders in satellite manufacturing) be considered as identity forming? Today it is certainly not a factor, but it could be built into one.

Man in space: Europe has essentially kept clear of the man-in-space programmes by maintaining a low-key participation (8%) in the ISS. Today three problems dominate the situation.

Access to the ISS: Europe has neither the means nor the will to replace the shuttle. The U.S. talks of using commercial launchers at least for cargo transportation. The human-rating of Ariane 5 is a complex matter that in the mind of the average citizen may appear to be necessary for establishing space. It is a fact that Ariane is the only European space product that has reached the status of identity forming (and still retains some of it), but it is considered more French than European. No such venture is contemplated in the next future. Orbiting men with Ariane-5 would have a limited impact (to be discussed...).

The ISS is so useless that the U.S., which paid 100 billion U.S. dollars to build it, wants to quit in 2015 (maybe 2020). Astronaut Reiter showed us the tiny Columbus (half the size of the Japanese module), lost on the majestic frame of the station. There is nothing in the ISS for European identity. Man in low Earth orbit is a lost cause in Europe.

The Moon: There is a strong possibility that manned missions to the Moon will be led by the U.S. and China. If this possibility materialises, I believe it should not be conceived as a race like Apollo, but as an Apollo of the 21st century, with a new paradigm: not going to the Moon, but going to the Moon together. Europe has the opportunity to propose its model of cooperation among equals, following the examples of CERN or ESA, and could play the role of honest broker in its conception and of facilitator in the implementation of an international Lunar Base that would not be a base but a multiple ensemble of various establishments. The European contribution would not be limited to the task of a facilitator but the hardware to be provided to the programme and the part played in operations including astronauts will be the subject of reflection at the Ministerial level in 2010.

Exploration beyond the Moon: After a very slow start, ESA robotic missions to planets have emerged as scientific successes. Have they contributed to European identity, as the NASA Hubble telescope has enhanced the image of American science? Definitely no, and this question will be briefly reviewed later. A robust scientific programme is necessary for Europe, and it is up to the science community and to agencies to organise an intelligent, permanent and well planned outreach. It is in this scope that Europe should adopt a project of Martian sample return as a flagship mission either alone (preferred option) or in cooperation with NASA and other partners (realistic option). A manned mission to Mars is not a subject for today.

The image of European space: One of the major factors in the absence of European space from the consciousness of the European public is the domination of the American cultural machine. This is not only NASA but essentially Hollywood, its TV productions and movies, and its extension – the Internet. As an anecdote, it was reported that all toy models of space objects available in European shops are American. The relationship of the space agencies and of the science and technology community with the media and the education system has to be built on new bases, so as to enable the fabrication of myths and the possibility to coin reality needed by Europe to exist in the minds. A strategic long term communication policy has to be conceived and implemented by high level professionals. Europe has a story to tell but that story has not yet been told. No Prince.

17 November 2009

Appendix

The ESA Columbus Essay Competition: "The value of human spaceflight for European citizens", November 2007

A.1.1 The value of human spaceflight for European citizens – written as a rhyme (first prize)

Daniela Petrova

What is the value of the human
Spaceflight for the European?
Some people perhaps feel detached
From the human spaceflight touch,
Perhaps because they have not been
To the heights beyond our atmospheric screen,
Perhaps because they have not touched
A rocket or its launching pad,
Perhaps because they have never thought
How much we owe to this field wrought.
Yet, despite it all, each and every one of us
Has benefited from this research thus
Lets remind ourselves in brief
Why we value the human spaceflight leitmotif.

There are benefits innumerable: the societal,
Multi-national, cultural, environmental…
Research, health and education,
Technology, business and innovation.
We value all the things
That would eliminate poverty, disease
And lead to world peace.

So when we look so closely,
At the common values we hold mostly,
It is possible to see revealed
That all our actions in this field
Have led to the improvement rife
For better living standards and quality of life.

Before recounting in more detail
There is a drop of history to feel.
Without the numerous inventors
That have given us the transportation by land, sea and air favours,
The engines and mechanics
On which such as Karl Benz and Felix du Temple worked despite the sceptics;
If it was not for the aspiration to the Moon
In Jules Verne's De la Terre à la Lune
Or for Beethoven's Moonlight Sonata's passion,
Then where would be our inspiration?!
Following our lengthy history
Of exploration and glory
In the post-Apollo mission,
Which made reality out of fiction,
ESA significantly contributed
In Spacelab and subsequently distributed
Work to many an European astronaut
From Thomas Reiter to Ulf Merbold
Moulding the European future
As a leading human spaceflight figure.
Today, with permanent human occupation
Aboard the International Space Station
We have had, still do and will, the opportunity
To utilise the unique conditions of microgravity.

Now to mention more specifically,
The value for our society and industry.
The European Commission rightfully recognised the potential
For economic growth, jobs and industry as essential
In the contribution to international cooperation,
Political and industrial cohesion, and multicultural integration.
With increasing international coordination
And growing peace between nations,

There will be military resource translocation
From the army towards peaceful space exploration.
Human spaceflight has an important action
In all of the above and education.
Not only in scientific principle demonstration
But also in fundamental inspiration,
Offering the future generation
A role-model for aspiration.
Consequently, there have been developed
Education projects that have enveloped
European students in science,
Mathematics, languages and art brilliance
As well as parabolic flights
That will lift them to new heights.
After many decades
Of human spaceflight experience,
We have moved beyond adventure
To acting as a technological aider.

Extraterrestrial research has allowed us to explore
Theories of relativity and origins of life closer.
Thanks to Earth observation and successful navigation
Disasters are averted between planes, ships and transportation.
Weather prediction and communication
Have allowed emergency evacuation,
Avoiding the consequences
Of natural disasters and loses.
Laser-cooled atomic clock developments
Allow for new, ultra precise measurements
While studies in Interactions
Between Cosmic and Atmospheric Particle Systems
Allow better understanding of pollutants and cloud formation
For future successful manipulation.
Astronauts have played a crucial part to date
To observe, interpret, adjust and update
Experiments for a successful mission,
Which with only robots would go without completion.

Human space exploration
Also provides a unique occasion

For testing interactions
Between robotics and humans.
The European Robotic Arm marks a new ERA
Of amplified human strength with a new gear.
The extreme space environment
Has led to the development
Of many high performance materials
For golf clubs, artificial hips and Shuttles,
High in insulation, durability
Corrosion and weight stability.
Further valuable spin-offs being
In construction and engineering,
Providing better insulation
Air conditioning and water purification,
Which is as readily applicable to business centres
As to Third World countries and remote sectors.

Increasing environmental awareness
Has led to questioning of process and resource
Utilisation, so that the closed loop system
Has become a primary concern.
Human spaceflight research
Has led to as much
As seventy percent success
In recycling excess.
Regeneration of air and clean water,
With some food production (even better),
Will be applicable in any
Extreme or closed environment for many
Whether in remote Antarctica, Africa, Sahara
Further than Mars, in a submarine, plane or tundra.

In addition, human spaceflight
Has helped develop new medicines to fight
Negative physiologic adaptations
Both in space and for Earth's millions.
Portable and automated devices
Lead to new diagnostic choices
For cardiovascular disease
To prevent its incipient malice;

The ESA Flywheel Exercise Device
Offers new rehabilitation and athletics exercise,
While Euromir-95 mission's 'Osteospace'
Offers a safe ultrasound alternative to X-rays;
Ophthalmological and neurological examinations
Have benefited from the VOG and 3D-eye tracker detection;
Respiratory Gas Analysers
Efficiently predict dysfunctions,
And ultraviolet protection is now available
For children with Xerodema Pigmentosum,
While the Mamagoose suite preserves the breath
Of babies from cot death.
Such remarkable scientific advances,
Due to human spaceflight rises,
Have led to increasing numbers
Of European centenarians.
Like a feed-forward reaction,
Human spaceflight attraction
Returns its investment many times over
By preventing disease or early death after.
The saved money in return
Can be used for a further space concern,
So it becomes cheaper and cheaper
To carry out missions in our favour.

The value of human spaceflight until now is unquestionable
And with Columbus' laboratory it will grow increasingly valuable.
Beyond the low Earth orbit
The future holds a further field,
That of explorations,
Of new worlds and destinations:
Initially to the Moon, Mars and Europa
Then to satellites and planets much further.
For there are many promises in the lands around us
Ice and water, stones and precious dust,
Like Helium-3, which will likely provide,
In future, clean and efficient energy might.

Our planet is like a rocky island
Suspended by an invisible hand

Into an ocean of space
Within which there are paths to trace.
Like Columbus and many others,
Who ventured across unprecedented expanses
And braved unknown perils
So that we may conquer the riddles,
Like the tamers of the land and sea
We tame the values in air and space we see!

A.1.2 The value of human spaceflight for European citizens (second prize)

Benjamin Lenoir, Ecole Polytechnique

"That one small step for a man, one giant leap for mankind." In 1969, less than ten years after Gagarin's flight in orbit, nobody questioned the relevance of human spaceflights. They were the goal of the whole process engaged in the two blocks: be the first to send a man in space. But, with the fall of the wall, a new area began, where there was no more reasons to show its technical superiority and therefore to send men in space. Then, considering the cost of such programs, the question of the value of human spaceflights came. And, today, the answer is obvious: despite the International Space Station, the majority of the space industry relies on unmanned missions.

However, if the economic rationality supports automatic spaceflights, human missions have not yet disappeared. The president of the USA even wants to return on to the moon and is peering at Mars. For a French citizen, member of the European Union, this asks the question of the value of human spaceflight for Europe and its citizens.

Space embodies the modern Frontier. But contrary to the pioneers in America, the conquest of space relies on technical capacities. Manned missions are therefore a wonderful way to energise research: it combines the dream of discovering new territories with the necessity to carry out researches to allow these trips in space.

In a world where innovation is the only way to stay at the top, the American success in the economic race is explained basically because of the strength of its research. The new French President understood this fact and aims at giving a new lease of live to research. But it would be a mistake not to think on a European scale: between the USA and China, European countries can be competitive only together. This is why human spaceflight is valuable for European citizens: it can be a strong bound on a European scale between many fields of research, which would directly benefit them because innovation is the key of economic competitiveness. From this point of view, human missions are much more interesting because they require much more technologies than unmanned missions and consequently have a broader impact.

Coupled with this capacity of human spaceflight to create innovation, manned mission is also a great way to conduct research: that is the very purpose of the ISS. When the Columbus laboratory will be attached to the ISS, Europe will have the capacity to explore new fields of science thanks to the unique experimental conditions in orbit. Having a strong human spaceflight capacity is consequently the assurance for Europe to conduct independently the experiments it wishes to conduct.

Despite the fact that research may seem very far from everyday life in Europe, there is no question about the fact that it enhances the competitiveness of Europe's production system and leads to a higher standard of living for every single European citizen.

Human spaceflight is valuable because of the research it induces and the economic benefits it brings. But the most important thing is somewhere else. Let's remember the reasons why Gagarin went in space and Armstrong walked on the moon. It was basically because of the national pride of two nations which wanted to lead the space exploration. And this period of space exploration certainly created a true national identity in these two countries.

In the last few years Europe seemed to have reached an impasse. As it worked in the USA or in Russia during the Cold War, human spaceflight could be a way to create the feeling of a European citizenship among the inhabitants of Europe, feeling which still lacks to build a strong European Union. If well advertised, space conquest can make people proud of Europe and therefore willing to build it. We, the European citizens, often hear speeches saying that we abide by the American way of thinking. This statement is certainly true concerning human spaceflight since Europe is not able to send a man in space by itself. That is why building an ambitious European program may create a keen interest among European citizens.

In addition to the creation of a European spirit, mastering human spaceflight is undoubtedly empowering Europe on the world stage. On the one hand, the European leaders would be able to speak of Europe as a whole. On the other hand, since space industry is very close to defense, mastering human spaceflight shows to the rest of the world the technical power of Europe. That is, for example, the reason why China is so keen on sending men into space: to show the rest of the world that they can do it and at the same time to say that they have the technical skills to develop missiles.

As a result, human spaceflight has major impacts for European citizens: it enhances their security while it helps creating a strong European Union.

In conclusion, despite the fact that most European citizens think they are not affected by human spaceflight, it has effects on their everyday life. In economic terms, it leads to more innovation in which the essential factor ought to be competitive. The consequence is a higher standard of living for the European citizens. But the human spaceflight adventure may also result, as it happened in the USA or in Russia, in the creation of a European spirit which would give Europe a stronger position on the world stage to defend its citizens' interest.

A.1.3 Looking over the edge – the value of human spaceflight for European citizens (third prize)

Regina Peldszus, Design Research Centre, Astronautics and Space Systems Group, Kingston University London

What happens to a pair of chopsticks when the parameter of gravity is changed? This is something science officer Don Pettit recorded during dinner preparations on ISS in 2003.[1] Casually picking up stray water spheres using the utensils after drinking from a bag, he demonstrated that in microgravity, liquids could be reliably picked up by chopsticks. This new feature completely changes the characteristics and utilisation potential of the tool. If it was not for Pettit's hands-on approach we might have never thought about this or expected it to be feasible from a non-orbital perspective, even if we were one of the hundreds of millions of skilled chopstick users in the world.

A tiny story about the changed potential of two simple rods in orbit can make us think outside the box'. They perhaps encourage us to anticipate the changing potentials of other objects, ideas, or people that the dangerous, beautiful and mundane idiosyncrasies of living in space might expose. From a designer interaction point of view, human spaceflight offers a vast chance to play, experiment and problem-solve. This form of resourcefulness, based on the ability to imagine, anticipate and improvise, is also what ground-based policy makers, civil engineers, scientists, designers, doctors and many others need to shape, sustain and possibly enrich the lives of their fellow citizens in a highly complex and often complicated world. For those not directly involved in the space programme – the majority of the over 500 million Europeans – the creative potential of spaceflight can inspire to look at the own environment in a different light. Supported by space-based infrastructure such as communications networks, these insights can be translated into everyday life.

Spaceflight has been changing, and will continue to change, the way we look at our world. This need not make us jaded or lose surprise over the quotidian. In fact, it can make us more accepting, curious and open-minded. By seeing things we take for granted from a different perspective, we acquire the chance to abandon set ways and prejudices. Like explorers who attempt new routes not merely out of necessity but because they are enticing, we depend on this degree of curiosity and the urge it gives us to move on, to see something else, and to re-evaluate and improve the status quo. This form of peaceful progress can only be achieved by critical, constructive and creative engagement with the present, and we need to cultivate such traits in a world where the new and unfamiliar is sometimes judged and rejected.

When there is no effort made to find out about who or what is behind the garden fence, the country border, the edge, there would be no information, no bonds, no new opportunities. If the states of Europe had not looked over their own edge with curiosity and acceptance, there would be no union today, and no joint space programme. Pulling together across boundaries is not only the modus operandi of human spaceflight in Europe today, but was also one of the pre-requisites when the ten founding states got together in 1975 to merge into what ESA is with its 18 members today.

Sending human individuals into space who both analyse and synthesise, and who relate their experience to billions of earthbound citizens, is one of the greatest opportunities to expand our own earthbound horizons on a global scale. Not just by creating new opportunities in science, technology, medicine or industry, but by triggering thought and perhaps setting off sparks of inspiration amongst people with different values, backgrounds and ambitions.

In a modern Europe made up of 27 countries that have each looked over their own borders for more than half a century in order to co-exist, ESA's intra-organisational work and its role as partner of global space projects plays an integral part. Far from being an idealistic vision, Columbus is a tangible manifestation of European collaboration, and a hardware example taking its place in the largest peaceful international project, ISS. Space activity is a vehicle for transcending boundaries in the scientific, political, economical and cultural framework of Europe and the world. The insights of human spaceflight, whether related to the big picture or incisive on a micro-scale, encourage us to look over the edge. Not only the edge of our earthly atmosphere, but the edge in our heads.

A.1.4 The value of human spaceflight for European citizens (fifth prize)*

Philippas Molyneux

In 1957, scientists from the Soviet Union successfully launched the worlds first artificial satellite, signalling the start of a highly competitive "space race" between the Soviets and their counterparts in the United States. Over the following decades, both countries fought to prove their scientific and military expertise by establishing a human presence in space. This fierce rivalry fostered a huge public interest in human spaceflight, as people all over the world were inspired by images of cosmonauts and astronauts orbiting the Earth and eventually standing on the moon. But this enthusiasm did not last. Humans have not set foot on the moon since 1972, and while the International Space Station now provides a base for different nations to experiment in space, the manned missions to explore other bodies in our solar system that many thought inevitable after the success of the Apollo programme have not been realised.

Fifty years after the launch of Sputnik kicked off the space age, the idea of human space exploration is once again beginning to gain popularity. More countries now have the technology and resources to contribute to this next step in spaceflight history. In 2003, China became the third nation to launch an astronaut into orbit, while Europe has increased its presence on the ISS with the addition of the ESA Columbus laboratory. India[2] and Iran[3] have both recently announced plans to develop human spaceflight capabilities, and in 2004, US President George W. Bush announced a goal of returning humans to the moon by 2020, leading to the development of NASA's Constellation Program. Although much of Constellation has been scrapped under the Obama administration, the new President has forecasted a manned mission to orbit Mars by the mid 2030s, with a Mars landing to follow.[4] Meanwhile, ESA's Aurora exploration programme will culminate with a manned mission to Mars by 2050.[5] This ambitious project will require collaboration with space agencies from around the world, but what can European citizens expect to gain from the work?

The European economy is strong, but since the industrial revolution it has been centred largely on manufacturing goods for trade and export. These days it is cheaper to produce many items outside of the European Union, and as a result Europe's economic growth has slowed. If the economy of the European Union is to stay strong, it must move away from the old industrial economy, and towards a more 'knowledge-based' economy in which scientific and technical expertise are more valuable than industry. Developments in human spaceflight would undoubtedly contribute hugely to this aim.

It may be argued that robotic space missions could fuel the new economy as successfully as manned missions, without the risks inherent to human exploration and at lower cost. However, the benefits of manned operations should not be underestimated. One of the main problems Europe faces in establishing a knowledge economy is that fewer people are choosing to study science and technology subjects to degree level- a trend that must be reversed to provide the well-educated and skilled workers necessary for such an economy to succeed. Human spaceflight may be one way to inspire more children to study science, both directly, through meeting astronauts and hearing about their work, and indirectly, through films and books inspired by the possibilities of space exploration.

While astronauts of all nationalities have the ability to encourage and motivate students, European children will relate more easily to European astronauts and so these would arguably have a bigger influence on the pupils' opinions of science. The scientific rewards of manned missions may also exceed those of unmanned missions. Robots lack the flexibility and problem-solving abilities of humans. A human in space may notice something interesting that robots would not know to look for, potentially altering the course of the mission, or saving a mission that might otherwise be unsuccessful. Humans are also able to cover much more ground in a given time: the inability of robot rovers to independently determine the best path to take means they must frequently send data about their position to Earth for analysis or spend long periods of time observing their surroundings in order to decide which way to go. The difference in efficiency between human and robot explorers can be quantified by comparing two missions: the Apollo 17 moon landing and the Mars Exploration Rover mission. In the 22 hours they spent exploring the lunar surface, the Apollo 17 astronauts covered 30.5 km and collected 110 kg of moon rock,[6] while after three years on Mars, the Spirit and Opportunity rovers had covered about 7 km[7] and 10 km[8] respectively.

Human spaceflight would clearly bring advances to planetary and space sciences, as well as in engineering and computing, as suitable space vehicles and communications systems would have to be designed or adapted for each mission. However, many of these advancements might well be made whether men or robots were sent to explore, albeit on different timescales. The main area of research in which manned missions would be significantly more useful than unmanned missions is biology. A human spaceflight programme would provide a unique opportunity to study the reaction of the human body to extreme conditions – specifically microgravity and high radiation levels – in space, potentially providing valuable insights into the workings of the body's organs and systems that could affect the way in which certain diseases are treated on Earth. For example, NASA works closely with America's National Institute of Health (NIH) to study the similarities between changes experienced by a human body in space and the aging process,[9] increasing our understanding of the problems people might face as they grow older.

The scope for progress in such a broad range of disciplines is one of the strongest arguments for human spaceflight in Europe. At a time when the European economy is changing to support more innovation in scientific and technological research, the potential of human spaceflight to advance science and inspire future researchers makes it a unique and exciting area of development. As more nations around the world start to develop the capability for space exploration, it is vital for Europe to maintain its position as a strong contributor to this field.

[*]Editorial note: no permission for reprint of fourth prize essay.

[1] Pettit, D. R. (2003). Expedition Six: Opening a Can of Honey. [video] March 15, 2003. Retrieved January 21, 2007, from *http://spaceflight.nasa.gov/gallery/video/station/expedition6/qtlow/waterscience.mov.*

[2] BBC News website. "India announces first manned space mission" retrieved 11 October 2010. *http://news.bbc.co.uk/1/hi/world/south_asia/8483787.stm.*

[3] BBC News website. "Iran aims to send man into space by 2019" retrieved 11 October 2010. *http://www.bbc.co.uk/news/world-middle-east-10747390.*

[4] NASA website. "Remarks by the President on space exploration in the 21st Century" retrieved 11 October 2010. *http://www.nasa.gov/news/media/trans/obama_ksc_trans.html.*

[5] ESA website. "Space Exploration Strategy – Related ESA programmes" retrieved 11 October 2010. *http://www.esa.int/SPECIALS/Space_Exploration_Strategy/SEMA9M0YUFF_0.html.*

[6] Kennedy Space Center website. "Apollo information- Apollo 17" retrieved 11 October 2010. *http://www-pao.ksc.nasa.gov/history/apollo/apollo-17/apollo-17.htm.*

[7] National Space Science Data Center website. "Spirit" retrieved 11 October 2010. *http://nssdc.gsfc.nasa.gov/nmc/masterCatalog.do?sc=2003-027A.*

[8] National Space Science Data Center website. "Opportunity" retrieved 11 October 2010. *http://nssdc.gsfc.nasa.gov/nmc/masterCatalog.do?sc=2003-032A.*

[9] NASA Human Spaceflight website. "STS-95 payloads – Human Research" retrieved 11 October 2010. *http://spaceflight.nasa.gov/shuttle/archives/sts-95/cargo/factsheets/index.html.*

A.2 Timeline of the construction of Europe after 1945

1946

September

19

Winston Churchill calls for a "kind of United States of Europe" in a speech he gives at the Zurich University.

December

17

The European Federalist Union is set into place in Paris, France.

1947

June

3

U.S. Secretary of State George C. Marshall proposes the Marshall Plan, or European Recovery Plan (ERP), to all the countries of Europe.

July

17

The Congress of the Committee for the Co-ordination of the European Movements takes place in Paris, incorporating several different European movements.

1948

March

17

The Western Union Treaty (Brussels Treaty) is signed by Belgium, France, Luxembourg, the Netherlands and the United Kingdom.

April

16

The Treaty of Paris is signed, creating the Organisation for European Economic Co-operation (OEEC) to help administer the Marshall Plan.

May

7–11

Organised by the International Committee of the Movements for European Unity, the Hague Congress gathers 750 delegates from around Europe. It intends to discuss ideas about the development of European political cooperation.

October

25

The European Movement is created.

1949

May

5

The Treaty of London is signed by ten states, creating the Council of Europe.

August

8

First session of the Council of Europe in Strasbourg.

1950

May

9

In a speech inspired by Jean Monnet, the French Foreign Minister Robert Schuman proposes integrating the coal and steel industries of Western Europe.

June

3

Belgium, France, Luxembourg, Italy, the Netherlands, and Germany subscribe to the Schuman declaration.

August

26–28

The Council of Europe Assembly approves the Schuman plan.

September

19

Creation of the European Payments Union.

November

4

Signature in Rome of the Convention for the Protection of Human Rights and Fundamental Freedoms, drafted by the Council of Europe.

1951

February

15

A meeting in view of the creation of the European Defence Community is held in Paris, France, Belgium, France, Italy, Luxembourg and Germany attend the meeting alongside six observer countries (USA, Canada, Denmark, Norway, The United Kingdom and The Netherlands).

April

18

The Six (Belgium, France, Germany, Italy, Luxembourg, Netherlands) sign the Treaty of Paris establishing the European Coal and Steel Community (ECSC).

1952

May

27

The Six (Belgium, France, Germany, Italy, Luxembourg, Netherlands) sign in Paris the European Defence Community (EDC) Treaty.

1953

March

7

The procedure regulation of the European Court of Justice is published in the Official Journal of the ECSC. As of today appeals foreseen by the Paris Treaty can be placed to the Court.

9

Paul-Henri Spaak, President of the ad hoc Assembly created on September 10, 1952, hands to G. Bidault, President of the ECSC Council, a draft treaty instituting a political European Community. Such Community would aim at safeguarding human rights and fundamental rights, at guaranteeing security of Member States against aggression, at ensuring the co-ordination of Member

States' external policy and at progressively establishing the Common Market. Five institutions are foreseen in the draft treaty: a European Executive Council, a two-chamber Parliament, a Council of National Ministers, a Court of Justice and a Economic and Social Committee.

1954

August

30
The French National Assembly rejects the European Defence Community.

October

23
Following the London Conference, agreements on a modified Brussels Treaty are signed in Paris, and the Western European Union (WEU) comes into being.

1955

June

1–3
Messina Conference attended by the Six to discuss a memorandum presented by the Benelux countries. The Ministers agree to consider the extension of European integration to all sectors of the economy. A committee of experts chaired by Paul-Henri Spaak is given the task of drawing up a detailed report on the feasibility of a general economic union and a nuclear energy union (the Spaak Report).

December

8
The Council of Ministers of the Council of Europe adopts as it emblem the blue flag hosting 12 golden stars.

1956

May

29–30
The Foreign Ministers of the Six meet in Venice and adopt the Spaak Report.

June

26
Negotiations for the drafting of the instituting texts for the EEC and Euratom open in Brussels.

1957

March

25

The Treaties establishing the European Economic Community (EEC) and the European Atomic Energy Community (Euratom) are signed by the Six (Belgium, France, Germany, Italy, Luxembourg, Netherlands) in Rome. They are referred to as the "Treaties of Rome".

1958

January

1

The Treaties of Rome enter into force. The EEC and Euratom Communities are set into place in Brussels. The Parliamentary Assembly, set up in Luxembourg, and the Court of Justice are common to all three Communities.

March

19

The session setting up the European Parliamentary Assembly is held in Strasbourg, France. Mr. Robert Schuman is elected President of the Assembly. This Assembly is to substitute the ECSC one.

July

3–12

Conference in Stresa, Italy to determine the guidelines of the future Common Agricultural Policy (CAP).

1959

January

1

The first steps are taken in the progressive abolition of custom duties and quotas within the EEC.

1960

January

4

The European Free Trade Association (EFTA) convention, regrouping Austria, Denmark, Norway, Portugal, Sweden, Switzerland and the United Kingdom is signed in Stockholm, Sweden.

May

11

The Council adopts the regulation relative to the European Social Fund. The aim of the Fund is promoting employment and geographical and professional mobility of workers within the Community.

1961

July

18

The Six meet in Bad Godesberg, Germany, and reiterate their intention to create a politically united Europe (the Bonn Declaration on the political Union).

31

Ireland formally applies to join the European Communities.

August

9

The United Kingdom formally applies to join the European Communities.

10

Denmark formally applies to join the European Communities.

September

1

The first regulation on free movement of workers comes into force.

1962

March

27–30

The Parliamentary Assembly decides to change its name into European Parliament.

April

17

Failure of the political Union (Fouchet Plan is rejected).

30

Norway formally applies to join the European Communities.

July

1

Customs duties on industrial products between member countries are reduced to 50% of their level of 1957.

1963

January

14

First veto of French President de Gaulle to the United Kingdom's application for membership.

22

Signature of the Elysée Treaty between Germany and France.

July

1

In pursuance of the decision to speed up the establishment of a custom union, the sixth intra-community reduction in customs duties and the second alignment on the common external tariff becomes effective.

September

23–24

The Council reaches an agreement in principle on the question of the merger of the Executives.

1964

July

1

The European Agriculture Guidance and Guarantee Fund (EAGGF) is set into place.

15

Costa/ENEL ruling. The European Court of Justice holds that Community law overrules national law.

September

18

The Council reaches an agreement as to the composition of the single Commission that will ultimately have nine members.

1965

April

8

The Treaty establishing a single Council and a single Commission of the European Communities is signed in Brussels.

July

1

Beginning of the "Empty Chair" crisis.

1966

January

28–30

The "Luxembourg compromise" is adopted, putting an end to the "Empty Chair" crisis. Member States can now invoke "vital national interests" to block majority decisions in the Council.

May

11

The Council adopts decisions and resolutions governing the timetable and the financing of the Common Agricultural Policy (CAP) and determines its objectives in other fields of common interest.

1967

May

11

The United Kingdom re-applies to join the Community. It is followed by Ireland and Denmark and, a little later, by Norway. General de Gaulle is still reluctant to accept British accession.

July

1

The Merger Treaty, fusing the Executives of the European Communities (ECSC, EEC, Euratom), enters into force. From now on the European Communities will have a single Commission and a single Council. However, both continue to act in accordance with the rules governing each of the Communities.

1968

July

1

Customs union enters into force. Remaining customs duties in intra-Community trade are abolished 18 months ahead of what was scheduled in the Rome Treaty and the Common Customs Tariff is introduced to replace national customs duties in trade with the rest of the world.

1969

July

16

The Commission submits to the Council a memorandum on replacing the financial contributions of the Member States by Communities' own resources and on increasing the budgetary powers of the European Parliament.

22–23

The Council resumes the examination of the application for EC membership of the United Kingdom, Denmark, Ireland and Norway. It asks the Commission to bring up to date the opinion it rendered on the matter in September 1967.

December

1–2

A Summit meeting is held in The Hague, the Netherlands. The heads of State or Government confirm their willingness to maintain the gradual advance towards a genuine economic and monetary union and the close alignment of social policies this entails, and reaffirm their agreement on the principle of the enlargement of the Community.

1970

March

4

The Commission submits to the Council a memorandum on the preparation of a plan for the establishment of economic and monetary union.

6

The Council assigns a committee of experts presided by Pierre Werner, to put out proposals for achieving economic and monetary union and entrusts a second committee of experts presided by Etienne Davignon, to issue proposals for political cooperation.

April

22

Signature of the Treaty of Luxembourg. The Council decides the gradual introduction of a system of own-resources under which the Community will receive all customs duties on products imported from non-member countries, all levies on agricultural imports and resources deriving from value-added tax. They also decide to extend the budgetary powers of the European Parliament.

October

17

Publication of the Werner report on the economic and monetary union.

27

Publication of the Davignon report on the European political cooperation.

1971

March

22

The Council adopts the Werner Plan to strengthen coordination of economic policies. The Member States have to take measures to harmonise their budgetary policies and to reduce the margins of fluctuation between their currencies.

31

European Agreement on Road Transport (AETR) ruling. The European Court of Justice develops principles that draw the line between the powers of the Community and those of the Member States. It makes clear that, as common rules are introduced, only the Community is in a position to take over and fulfil, in respect of the entire field of application of the Community legal order, obligations undertaken vis-à-vis non-member States.

1972

January

22

Signature of the Accession Treaties for the United Kingdom, Ireland, Denmark and Norway.

April

10

Creation of the European currency snake, allowing a maximum drift for national currencies of 2.25% around the fixed parity.

September

12

The Ministers of Finance of the six and of the four countries that applied for membership meet in Rome, Italy. They agree that in the first stage of economic and monetary union it is necessary to set up a European Monetary Cooperation Fund.

25

A referendum is held in Norway on the country joining the European Communities. The majority is unfavourable to accession.

October

2

A referendum is held in Denmark on the country joining the European Communities. The majority is in favour of accession.

9

Following the unfavourable vote of the referendum, the Norwegian Government declares that Norway will not bring the accession ratification Bill before the Parliament. The European Parliament passes a resolution containing definite suggestions and requests for the abolition of checks at intra-Community frontiers.

16

The United Kingdom ratifies the acts relating to the accession to the European Communities.

19–21

A Summit meeting is held in Paris, France. Heads of State or Government define new fields of Community action (concerning regional, environmental, social, energy and industrial policies) and reaffirm 1980 as the deadline for the achievement of economic and monetary union.

1973

January

1

Denmark, Ireland and the United Kingdom join the European Communities. The Community Free Trade Agreement with Austria, Switzerland, Portugal and Sweden comes into force.

December

14–15

A summit conference is held in Copenhagen, Denmark. The energy crisis leads the Member States to agree on the introduction of a common energy policy. A statement on the European identity, drafted as part of political cooperation arrangements and approved by Foreign Ministers, is released.

1974

December

9–10

Paris summit, where the Heads of State or Government of the Nine considered it essential to ensure progress and overall consistency in the activities of the Communities and in the work on political co-operation.

1975

February

28

First ACP-EEC Convention is signed in Lomé by the 9 Member States of the Community and their 46 partners of the African, Caribbean and Pacific (ACP) States (Lomé Agreements).

June

5

The outcome of the British referendum reveals that 67.2% of voters are in favour of the United Kingdom remaining a member of the Community.

12

Greece applies to join the Community.

August

1

The final Act of the Conference on Security and Cooperation in Europe is signed in Helsinki, Finland, by 35 States.

1976

January

7

Publication of the Tindemans report, laying down how the term "European Union" might be interpreted. It advocated consolidation of the existing institutions and the development of common policies.

March

4

Two Protocols with Malta are signed. The 1970 Association Agreement is extended to take in agriculture and cooperation.

September

20

Signature of the Act on the election of the European Parliament by universal suffrage.

1977

March

28

Portugal applies to join the Community.

June

1

The Treaty reinforcing the budgetary powers of the Parliament enters into force.

July

28

Spain applies to join the Community.

1978

April

3

The Community and the People's Republic of China sign a trade agreement, which comes into force on June 1.

September

26

The Council adopts some cooperation agreements with Algeria, Morocco, Tunisia, Egypt, Syria, Jordan and Lebanon and additional financial protocols with Israel, Portugal and Malta.

October

9

The representatives of the Governments of the Member States meeting within the Council signed the Convention on the Accession of Denmark, Ireland and the

United Kingdom to the Convention on Jurisdiction and the Enforcement of Judgments in Civil and Commercial Matters.

November

1
The Cooperation agreements with Maghreb countries (Algeria, Morocco, Tunisia) and the Mashreq countries (Egypt, Syria, Jordan, Lebanon) and additional financial protocols with Israel, Portugal and Malta enter into force.

December

4–5
A European Council is held in Brussels, Belgium. It establishes the European Monetary System based on a European currency unit (the ECU) and decides to call in a committee of leading personalities to consider adjustments to institutional mechanisms and procedures in the context of enlargement.

1979

March

13
The European Monetary System enters into force.

June

7–10
First election of the European Parliament by universal suffrage.

October

28
Second ACP-EEC Convention is signed in Lomé by the 9 Member States of the Community and their 59 partners of the African, Caribbean and Pacific (ACP) States.

1980

April

2
A cooperation Agreement signed between Community and Yugoslavia.

October

1
The EEC – Association of south-east Asian Nations (ASEAN) Cooperation Agreement comes into force.

1981

January

1

Greece becomes a Member State of the Community.

March

23

Signature of an agreement for commercial and economic cooperation between the Community and India.

1982

October

1

The framework agreement on cooperation between the Community and Brazil enters into force.

1983

March

21–22

A European Council is held in Brussels, Belgium. It confirms the priority objectives set out at the Copenhagen European Council of 1982, that is: the priority objectives set out for the economic and social fields and its political engagement in favour of the enlargement.

December

17

An Economic cooperation agreement between the Community and the Andean Pact countries is signed in Cartagena, Colombia.

1984

February

14

The draft Treaty on the establishment of the European Union (Spinelli draft) is passed by the European Parliament by a large majority.

28

The Council adopts decision on European strategic programme for Research and Development in information technology (Esprit).

June

7

The Council and the representatives of the governments of the Member States adopt a resolution on the reduction of border checks on people.

December

3–4

A European Council is held in Dublin, Ireland. It decides to reinforce the European Monetary System (EMS) and to grant the ECU a more important role.

8

Third ACP-EEC Convention is signed in Lomé by the 10 Member States of the Community and their 65 partners of the African, Caribbean and Pacific (ACP) States.

1985

June

14

The Schengen Agreement on the elimination of border controls is signed by Belgium, Germany, France, Luxembourg and the Netherlands in Schengen (Luxembourg).

December

2–4

A European Council is held in Luxembourg. The Ten agree to amend the Treaty of Rome and to revitalise the process of European integration by drawing up a Single European Act.

1986

January

1

Spain and Portugal become Member States of the Community.

February

17 and 28

The Single European Act modifying the Treaty of Rome is signed in Luxembourg and The Hague.

1987

May

26

Following the results of the referendum on the ratification of the Single European Act held in Ireland, the Irish constitution is amended in order to allow Ireland to ratify the Single Act and to deposit the instruments of ratification.

July

1

The Single European Act held in the Single Act comes into force.

1988

February

13

Adoption of the "Delors I Package" reforming the Community Budget, guaranteeing its financing and improving the annual budgetary procedure.

1989

April

12

Adoption of the Delors Report laying down three conditions for the fulfillment of the Economic and Monetary Union: full and irreversible convertibility of currencies, the establishment of the free movement of capital, the adoption of a single currency.

July

17

Austria applies to join the Community.

November

9

Fall of the Berlin Wall.

December

8–9

A European Council is held in Strasbourg, France. It decides to convene an Intergovernmental Conference before 1990 to draw up amendment to the Treaty for the final stages of economic and monetary union. Heads of State or Government of 11 Member States adopt the Charter of Fundamental Social Rights of Workers.

1990

April

28

A Special European Council is held in Dublin, Ireland. It agrees on a common approach on German unification and on the Community relations with Central and Eastern European countries.

June

19

The Schengen Agreement on the elimination of border checks is signed by the Benelux countries, France and Germany.

20

The EEC and EFTA start formal negotiations for the creation of the European Economic Area (EEA).

July

1

First step of the EMU with the introduction of complete freedom of movement of capital.

November

19–21

A meeting is held in Paris, France. Thirty-four Heads of State or Government of the Conference on Security and Cooperation in Europe (CSCE) sign a Charter for a new Europe.

27

Italy signs the Schengen agreement.

December

15

Parallel opening of two intergovernmental Conferences in Rome: one on future plans for the EMU, the other on the future Political Union.

1991

February

25

Dissolution of the Warsaw Pact.

July

1

Sweden applies to join the Community.

September

1

The fourth Lomé Convention, on the development of the relations between the Community and the Africa, Caribbean, Pacific (ACP) States, enters into force.

December

9–10

A European Council is held in Maastricht, The Netherlands. It reaches an agreement on the draft Treaty on the European Union.

16

"Europe Agreements" are signed with Poland, Hungary and Czechoslovakia.

1992

February

3

The Council adopts positive measures to help Croatia, Slovenia, the Yugoslav Republics of Bosnia-Herzegovina and Montenegro and the former Yugoslav Republic of Macedonia.

7

The Treaty on the European Union is signed in Maastricht by the Foreign and Finance Ministers of the Member States.

March

18

Finland applies to join the Community.

May

2

Signature of the Agreement on the European Economic Area (EEA) in Porto.

November

25

Norway applies to join the Community.

December

11–12

Adoption of the "Delors II Package" at the European Council in Edinburgh, continuing the reforms undertaken by the "Delors I Package".

1993

January

1

Achievement of the internal market.

November

1

The Maastricht Treaty enters into force.

1994

January

1

The Treaty on the EEA enters into force.

May

25

The board of Governors of the European Investment Bank establishes the European Investment Fund.

26–27

An inaugural conference for a Stability Pact for Central and Eastern Europe is held in Paris, France.

June

25

Signature of the Accession Treaty for Austria, Finland, Sweden and Norway.

October

10

A conference on Security and Cooperation in Europe (CSCE) opens in Budapest, Hungary.

1995

January

1

Austria, Finland and Sweden become Member States of the Community.

March

14

The Council and the Parliament sign the Socrates Programme in the field of education.

20–21

The Stability Pact for Central and Eastern Europe is signed and adopted in Paris.

26

The Schengen Agreements enter into force.

December

14

Signature of the Dayton Peace Agreements for ex-Yugoslavia in Paris.

15–16

A European Council is held in Madrid, Spain. It sets March 29, 1996 as the starting date for the Intergovernmental Conference and confirms the introduction of the single currency ("euro") for January 1st, 1999.

1996

March

29

Opening of an Intergovernmental Conference in Turin to negotiate a revision of the Maastricht Treaty.

1997

July

16

The Commission presents the "Agenda 2000 – for a stronger and wider Europe" and its opinions on the applications of 10 Central and Eastern European countries.

22

The Western European Union (WEU) Extra-ordinary Council adopts a declaration, to be annexed to the final Act of the Amsterdam Treaty, on its role and its relations with the EU and the Atlantic Alliance.

October

2

Signature of the Amsterdam Treaty.

December

12–13

The European Council meets in Luxembourg and takes the decisions needed to launch the enlargement process. It also adopts a resolution on economic policy co-ordination.

1998

March

31

Opening of the negotiations with Hungary, Poland, the Czech Republic, Slovenia, Estonia and Cyprus.

April

30

Peace agreement on Northern Ireland.

May

3

A special Council decides that 11 Member States satisfy conditions for adoption of the single currency on 1 January 1999. The Commission and the European Monetary Institute set out conditions for determination of the irrevocable conversion rates for the euro.

June

1

Establishment of the European Central Bank.

1999

January

1

Introduction of the euro.

June

3–4

The European Council meeting is held in Cologne, Germany. It adopts the first European Union common strategy, which concerns Russia, and declarations on

Kosovo and on the strengthening of European common foreign and security policy, and designates Mr Javier Solana Madariaga High Representative for the CFSP and Secretary-General of the Council.

28–29
The Rio Declaration is adopted at the close of the first Summit of heads of state and government of the European Union, countries of Latin America and the Caribbean. It stresses the importance of developing a strategic partnership between the two regions and sets out 55 priorities for action.

2000

November

8
The Commission delivers to the Council the overall report on enlargement, consisting of progress reports assessing the preparation of the candidate countries and an "Accession Partnership" proposal identifying the key issues Turkey must address before starting accession negotiations.

2001

February

26
Following the December 2000 European Council meeting held in Nice, France, a new Treaty amending the Treaty on European Union and the Treaties establishing the European Communities, is signed (Treaty of Nice).

2002

February

28
The euro becomes the sole currency within the twelve participating Member States, as the period of dual circulation comes to an end. The opening session of the Convention on the Future of Europe is held in Brussels.

October

9
The European Commission recommends the conclusion of accession negotiations by the end of the 2002 with the following countries: Cyprus, the Czech Republic, Estonia, Hungary, Latvia, Lithuania, Malta, Poland, the Slovak Republic and Slovenia. The Commission considers that these countries will be ready for EU membership from the beginning of 2004.

2003

February

1
The Treaty of Nice enters into force.

2004

May

1
The Accession Treaty enters into force and the European Union's biggest enlargement ever in terms of scope and diversity becomes a reality with 10 new countries – Cyprus, the Czech Republic, Estonia, Hungary, Latvia, Lithuania, Malta, Poland, the Slovak Republic, and Slovenia – representing all together more than 100 million citizens, joining the European Union.

5
The European Parliament proceeds to a formal vote of approval for the 10 new Commissioners from the 10 new member states during its last part-session of the current legislature. The Council of Ministers also formally appoints the ten new Commissioners.

2005

April

13
The European Parliament gives its approval for the entry of Romania and Bulgaria into the EU. Actual accessions are scheduled to take place in 2007.

2006

February

The European Parliament adopts, by a large majority, a first-reading report on legislation opening up the EU single market for services. The Services Directive, also known as the Bolkestein Directive, is a major issue for the European Union.

2007

January

1
The accession of Romania and Bulgaria completes the fifth enlargement of the EU that began in May 2004, raising the number of Member States to 27 and the population within the Union to 492.8 million inhabitants.

October

18–19

A Key Summit on the Reform Treaty takes place in Lisbon, at which a precise text is agreed for the new 'Lisbon Treaty' which will be formally signed by all European leaders in Lisbon on 13 December 2007.

2008

November

10

The Council launches an EU joint military action to improve maritime security off the Somali coast. Operation Atlanta comes as a result of an increasing number of attacks on vessels by pirates.

December

11–12

The European Council meeting in Brussels approves a European Economic Recovery Plan to fight the financial and economic crisis.

12

The EU welcomes Switzerland in its 'Schengen area', which allows people to travel between countries without border checks whilst ensuring increased police cooperation.

2009

June

18–19

EU summit in Brussels prepares the way for a second referendum on the Lisbon Treaty in Ireland.

December

1

The Lisbon Treaty enters into force.

10–11

At a meeting of the European Council held in Brussels the 'Stockholm Programme' is adopted for the years 2010–2014 aimed at further development of freedom, justice and security.

2010

March

26

At a European Council meeting in Brussels, EU leaders adopt Europe 2020 targets and all 16 eurozone countries back a plan to help Greece deal with its deficit.

June

17

At a European Council meeting in Brussels, EU leaders adopt a 10-year strategy for smart, sustainable and inclusive growth: Europe 2020. They also decide to open accession negotiations with Iceland.

2011

January

18

The first "European semester" gets under way – a six-monthly cycle of economic policy coordination between EU countries which is meant to help prevent economic crises like the one in 2008–10.

A.3 Eurobarometer Survey "Space activities of the European Union". Analytical Report (published October 2009)*

Complete document online available from http//extras.springer.com

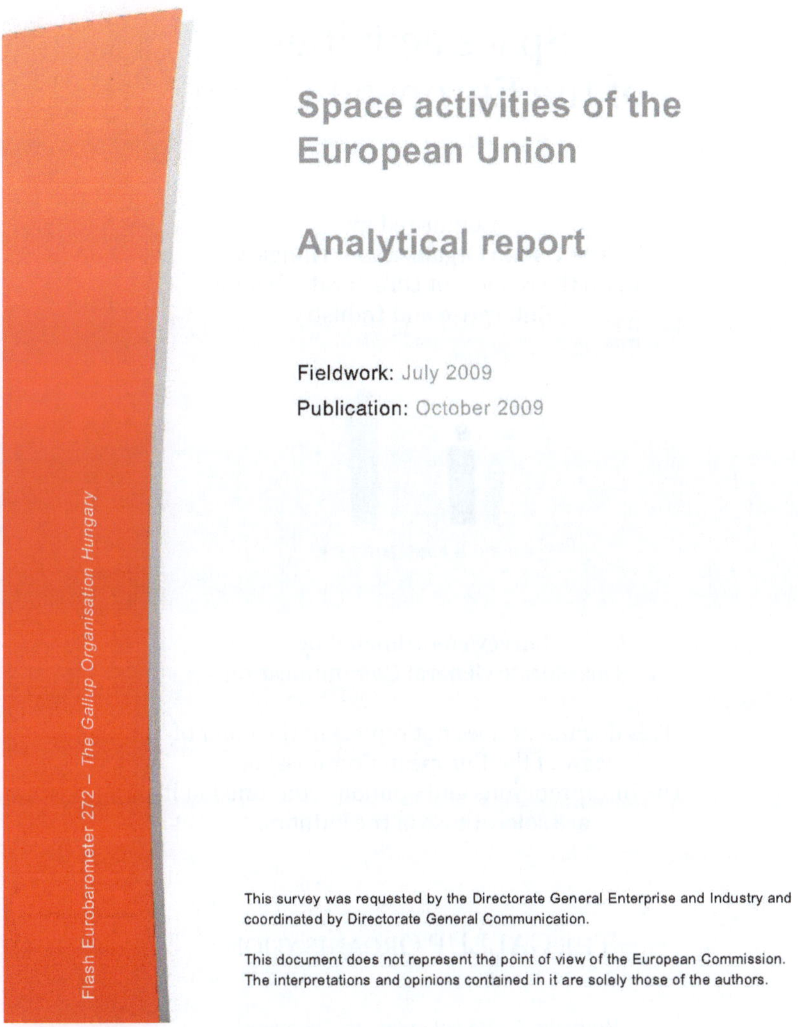

Flash Eurobarometer 272 – The Gallup Organisation Hungary

Space activities of the European Union

Analytical report

Fieldwork: July 2009

Publication: October 2009

This survey was requested by the Directorate General Enterprise and Industry and coordinated by Directorate General Communication.

This document does not represent the point of view of the European Commission. The interpretations and opinions contained in it are solely those of the authors.

---

*Source: European Union, Eurobarometer.
http://ec.europa.eu/public_opinion/flash/fl_272_en.pdf (pp. 1–19)
"The European Union does not endorse changes, if any, made to the original data and, in general terms to the original survey, and such changes are the sole responsibility of the author and not the EU"

Flash EB Series #272

Space activities of the European Union

Conducted by
The Gallup Organisation, Hungary
upon the request of Directorate General
Enterprise and Industry

EUROBAROMETER

Survey co-ordinated by
Directorate General Communication

This document does not represent the point of view of the European Commission.
The interpretations and opinions contained in it are solely those of the authors.

THE GALLUP ORGANISATION

Web-site for downloading the document:
http://ec.europa.eu/public_opinion/index_en.htm

Introduction

Space systems and space-based technologies are a critical part of the daily life of all European citizens and businesses. From telecommunications to television, weather forecasting to global financial systems, most of the key services that we all take for granted in the modern world depend on space to function properly. Research and development activities are co-ordinated within the framework of the overall European Space Policy, complementing the efforts of Member States and of other key players, including the European Space Agency.

Space systems are clearly strategic assets that demonstrate independence and an ability to assume global responsibilities. To maximise the benefits and opportunities that they can provide to Europe now and in the future, it is important to have an active co-ordinated strategy and a comprehensive European Space Policy.[1]

The Flash Eurobarometer *Space activities of the European Union* (Flash N° 272) was conducted in order to examine EU citizens' opinions and to assess: a) their awareness of space activities of Europe and the European Union, b) their perception of these activities, and c) their general attitude toward space exploration.

The survey fieldwork was conducted between 3 and 7 July 2009. Over 25,000 randomly selected citizens aged 15 years and over were interviewed in the 27 EU Member States. Interviews were predominantly carried out by fixed-line telephone, reaching ca. 1,000 EU citizens in each country (the size of the sample was 500 in Luxembourg, Malta and Cyprus).

Although interviews were predominantly carried out by telephone via fixed-lines, interviews were also conducted via mobile telephones and by face-to-face (F2F) interviews as appropriate. This methodology ensures that results are representative of the EU27 Member State population. In most of the countries where a large share of residents could not be contacted by fixed-line telephones (as many do not subscribe to such service), a mixed-mode methodology was employed to ensure that these individuals were questioned and this was done either through F2F interviews or by including mobile telephones in the sampling frame. For this survey, mobile telephone interviews were conducted in Austria, Belgium, Finland, Italy, Portugal and Spain, and some F2F interviews in the Czech Republic, Lithuania, Hungary, Slovakia, Latvia, Romania, Estonia, Bulgaria and Poland.

To correct sampling disparities, a post-stratification weighting of the results was implemented, based on key socio-demographic variables. More details on the survey methodology are included in the Annex of this report.

Please note that due to rounding, the percentages shown in the charts and tables do not always add up exactly to the totals mentioned in the text.

[1] See: http://ec.europa.eu/enterprise/policies/space/index_en.htm

Main findings

- The majority of European Union citizens regard European space activities as important from the perspective of the EU's future global role: one in five citizens considered such activities *very* important (20%) and a further 43% felt that space activities are important.

- When asked about the importance of developing various space-based applications for Europe, EU respondents were most keen on (further) developing environmental/natural monitoring systems: 58% found this *very* important. Regarding other services the mood remained generally positive; the proportion considering these *very* important remains in the one-quarter to one-third range (satellite-based communications, positioning system and satellite-based monitoring to improve citizens' security)

- EU citizens generally acknowledged that there may be various benefits related to space exploration, (it may add to human knowledge, it may help to protect our planet, it may help to find new raw materials and energy sources and it may boost economy through technological innovations), and showed a widespread support to extend EU activities in space exploration.

- 26% of all Europeans thought that the EU should *definitely* do more in the field of space exploration, and 38% felt that it should *perhaps* put more emphasis on this field. 30% provided a negative response to this question (28% opposed more involvement and 2% indicated that it does not matter for them whether or not the EU is more active in space exploration).

- Generally speaking there is balance between EU citizens that – in these times of economic and financial crisis – support and do not support an increase in the EU budget devoted to space activities. The 20% who felt that more budgetary resources should be allocated to space activities were outweighed – although only by a slim margin – by those who felt that the EU should reduce such spending (23%). The plurality (43%) felt that the current budget should be maintained. Overall there is strong support for EU funding of space activities at its current level (43%) or at increased level (20%).

1. Strategic importance of space research in Europe

The majority of European Union citizens regard European space activities as important from the perspective of the EU's future global role: one in five citizens considered such activities *very* important (20%) and a further 43% felt that space activities are important in this respect. In total, almost two-thirds of Europeans share the view that space activities are important for the future international position of the European Union.

About one in ten of those interviewed (9%) could not form an opinion in this matter, and 29% thought that such activities are not strategically important.

Importance of space activities for the future international position of the EU

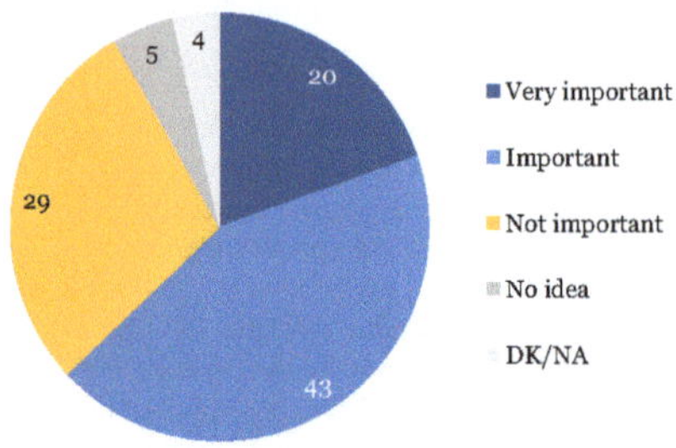

Q1. Do you think space exploration and other space activities are important for the future international position of the European Union?
Base: all respondents, % EU27

Those considering space activities as important were in the majority in each Member State of the EU. Even in those countries where the proportion of sceptical citizens was the highest (France: 41%, Austria: 39%, Denmark: 38%), most people believed that such activities were important for the EU's future global role.

The ratio of those considering such activities as "very important" remained at or below one-third of all interviewees, with the highest proportion in Bulgaria (33%), Greece (32%) and Ireland (32%). Still, in 22 Member States those who regarded European space activities important or very important exceeded 60%, signalling a positive mood behind initiatives in this area. The general appreciation (using an indicator that combines the "important" and "very important" replies) was highest in Cyprus (82%), Greece (80%) and Bulgaria (77%).

Importance of space activities for the future international position of the EU

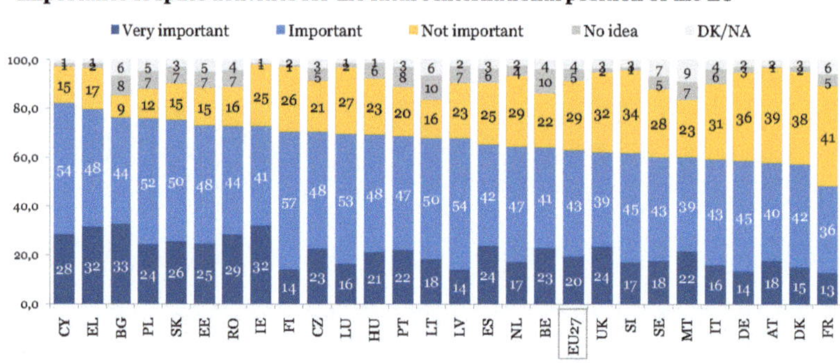

Q1. Do you think space exploration and other space activities are important for the future international position of the European Union?
Base: all respondents, % by country

Men (67% compared to 60% of women) and respondents from the youngest age group (15-24 years: 73%) were the most likely to think that space activities are strategically important for Europe, but in each socio-demographic segment, the overwhelming majority was supportive. The lowest support was detected among those with the least education: only 53% of those who completed their education at the age of 15 or earlier found space activities important, while 38% thought the opposite, see Annex Table 1b.[2]

1.1 Perceived benefits of space activities

The vast majority of Europeans agree that technology transfers from the space industry can contribute to innovation in terrestrial applications (24% agreed strongly and 50% to some extent). Only 16% expressed scepticism at this idea. In addition, almost two-thirds agreed that space activities may contribute to the success of the European economy at large: 16% agreed strongly and 48% agreed that space industry activity can boost European competitiveness, economic growth and create jobs in Europe. Pessimism regarding the latter aspect remained at 28% (almost 3 in 10 respondents).

Benefits of space technology for terrestrial applications

There is very little variation of opinions across Member States in the proportion of people believing in the potential of transferring space technologies to terrestrial applications; rather small minorities in each country of the EU expressed doubts in this regard.

Optimism was highest in the Netherlands (where 85% agreed either strongly or moderately that technologies derived from space activities contribute to developing innovative terrestrial applications), Finland (83%), Cyprus (81%) and Hungary (80%). Hungary had in fact the most people who *strongly* agreed that such transfers foster innovation (36% of all respondents).

While remaining largely supportive, the highest levels of scepticism were recorded in the Czech Republic (22%), Italy (21%) and Malta (20%).

Technologies derived from space activities contribute to developing innovative terrestrial applications

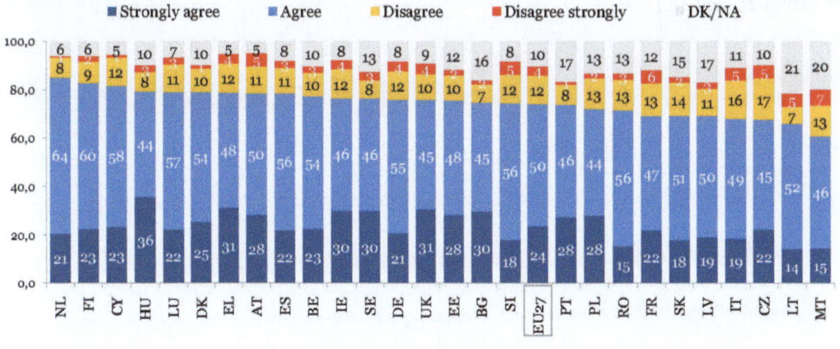

Q2. Please tell me how much do you agree or disagree with the following statements:
Base: all respondents, % by country

Benefits of space technology for technological and economic development

As indicated above, most EU citizens believe in the benefits of space technology for technological and economic development even if these seem somewhat less evident for Europeans than the technological advantages. (In 2006 the European space manufacturing industry employed around 29,000 people and generated a turnover of €5bn[3].). Despite an overall optimism in each Member State, the survey found a

[2] Please note that all socio-demographic analyses were carried out at EU27 level.

[3] http://ec.europa.eu/enterprise/policies/space/faq/index_en.htm

considerable minority in every country who disagreed with the statement that the space industry could contribute to economic growth, industrial competitiveness and job creation. (On average the proportion of sceptics in this regard is 28%, but in Slovenia: 43%, Czech Republic: 37% and Latvia: 35%).

Space activities contribute to industrial competiveness, growth and creation of jobs in the European Union

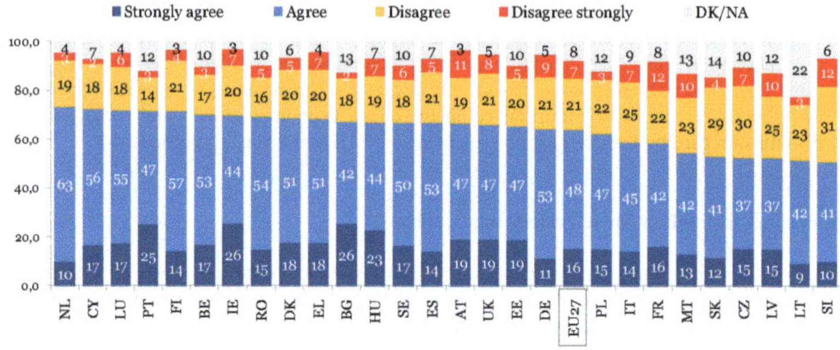

Q2. Please tell me how much do you agree or disagree with the following statements:
Base: all respondents, % by country

Still, everywhere, including in countries where the highest levels of scepticism were recorded, those who felt that a successful and strong space industry would be beneficial to the European economy outnumbered the sceptical. Public opinion was the most convinced of the existence of such economic benefits in the Netherlands, Cyprus (in both countries 73% agreed, in total), Luxembourg and Portugal (72% both).

Men, younger respondents, better educated and metropolitan citizens were more likely than others to agree that there were economic and technological benefits to be gained from space activities (see Annex Tables 2b. and 3b.).

2. Satellite-based services

European Union countries independently as well as the European Space Agency (ESA) operate a number of satellites to provide information for terrestrial applications (surface/ meteorological/ environmental monitoring, broadcasting and communication, positioning, security applications, etc.).

2.1 Familiarity

The majority of Europeans indicated that they are aware that such European Earth Observation satellites exist (56%), and 22% were also confident that they knew what these satellites are used for (34%, albeit aware of their existence, were not sure what function they fulfil). 42% were not aware of (at least the specifically "European") Earth Observation satellites.

Awareness of the European Earth Observation satellites

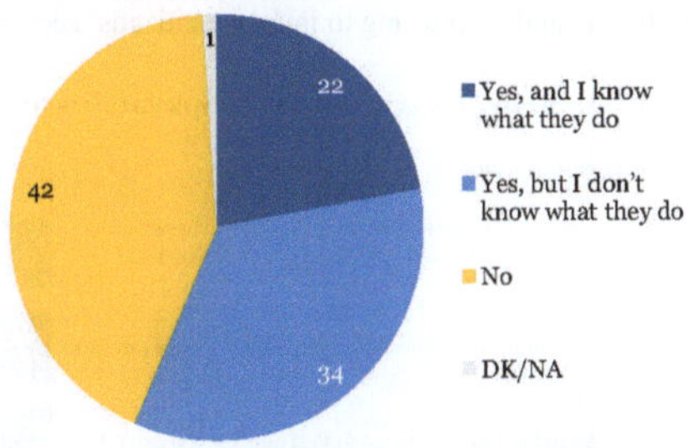

Q3. Have you heard about the European Earth Observation satellites?
Base: all respondents, % EU27

Awareness levels varied significantly across Member States, with respondents interviewed in Slovenia (of whom 76% were aware of European Earth Observation Satellites), Spain (72%), Romania (70%) expressing the highest ones.

On the other hand, a large number of respondents in the UK (73%), Sweden (72%), Ireland (70%) and Finland (66%) stated that they had never heard of European Earth Observation satellites. Generally, interviewees in Slovenia (34%), Spain (32%), France and Hungary (30% both) were most likely to indicate that they were familiar with the function of these satellites.

Awareness of the European Earth Observation satellites

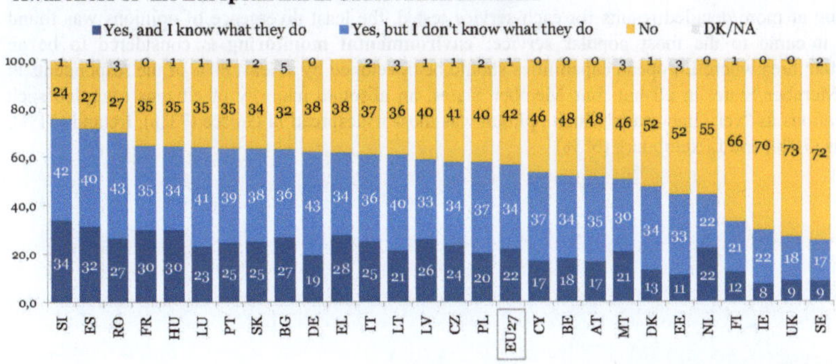

Q3. Have you heard about the European Earth Observation satellites?
Base: all respondents, % by country

Considering awareness of European Earth Observation satellites, the socio-demographic patterns changed. While men and the best educated were still the most aware of the satellites' existence, young people were the least informed of all age groups (54% of those aged 15-24 indicated that they had not heard about them, compared to a 42% EU average, see Annex Table 4b.).

2.2 Satellite-based applications

When asked about the importance of developing various space-based applications for Europe, EU respondents were most keen on (further) developing environmental/natural monitoring systems: 58% found it *very* important that Europe has observation systems able to monitor natural and environmental threats (a mere 6% found this unimportant). When it comes to other applications, the mood remained positive with about two-thirds finding these important or very important, but except for Earth observation systems, at European level those who consider the development of this applications "very important" was lower than those that responded that they were important. Other applications mentioned concerned communications (regarded as at least quite important by 75%), positioning system (67%) and monitoring to improve citizens' security (67%).

Development of various space applications for Europe

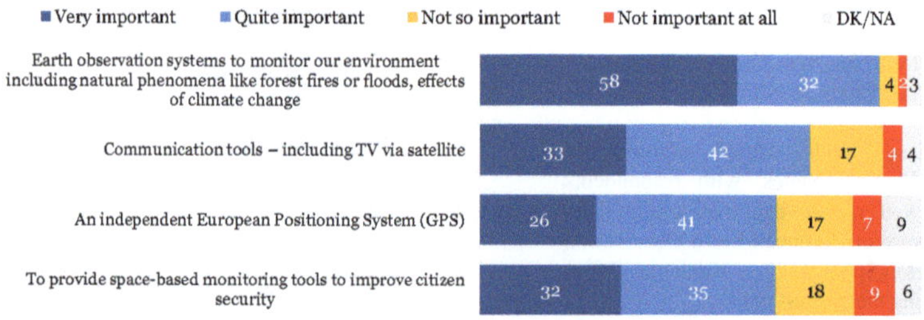

Q4. How important is in your view to develop the following space applications for Europe?
Base: all respondents, % EU27

Overall, 27% found it unimportant to develop satellite-based applications to improve citizens' security, and 24% had a similar opinion about an independent European positioning system. 21% did not think that European satellite-based communication services should be (further) developed.

Looking at more detailed results for each service tested, the least divergence of opinions was found when it came to the most popular service: **environmental monitoring** is considered to be an important area where European capabilities should be developed by at least 80% of the respondents in each Member State. In all but four Member States, an absolute majority of citizens regarded such applications as "very important". Such opinion was most widespread in Greece (73%), Ireland (71%), Cyprus (70%) and Luxembourg (68%).

Development of Earth observation systems to monitor our environment including natural phenomena like forest fires or floods, effects of climate change

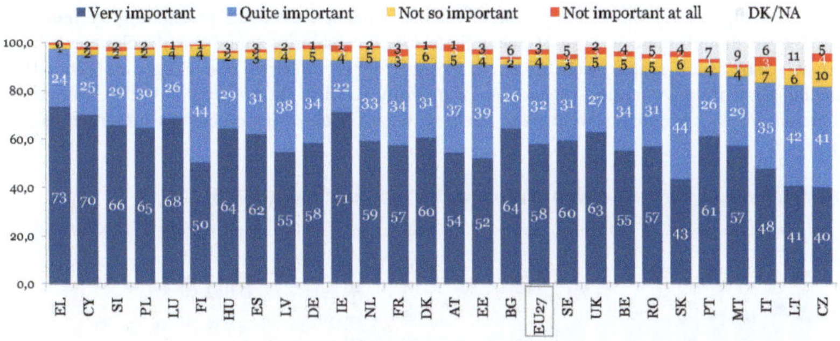

Q4. How important is in your view to develop the following space applications for Europe?
Base: all respondents, % by country

Reflecting the extremely widespread approval, those who regarded such development as important or very important varied only slightly among the various socio-demographic groups analysed. (See Annex, Table 6b.). The share of those supporting such developments varied between 89% and 93%, with the exception of those with the lowest education: ‘only’ 84% of those who finished school at the age of 15 or earlier thought that the development or improvement of such applications was important.

Such outright positive attitude was less frequent, although still overwhelmingly positive, when respondents were asked about **satellite-based communication services** (e.g. satellite TV): those who regarded such developments as very important made up an absolute majority only in three Member States (51% answered so in Ireland, Hungary and Bulgaria). Nevertheless, over 60% in each country found such developments at least quite important. Overall, the most affirmative replies were recorded in Cyprus (88% quite and very important), Ireland, Hungary (both 87%), and Poland (86%).

Development of communication tools – including TV via satellite

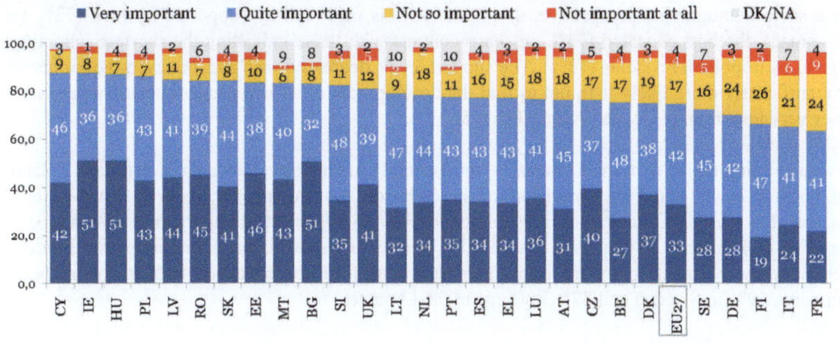

Q4. How important is in your view to develop the following space applications for Europe?
Base: all respondents, % by country

Countries where a relatively large proportion of respondent do not think that it is important to develop space applications in Europe related to communication services include some of the largest Member States. In France 33%, in Finland 31% in Germany and Italy 27% saw the development of space-based communication tools as not so important or not important at all.

Differences across the various socio-demographic segments remained slight, however men were clearly more likely to agree that such developments are important (78%) than women (71%), and those with degree-level education (having studied beyond the age of 20) were also more supportive (78%)

compared to those who completed their education at the age of 15 or earlier (67%). (See Annex, Table 7b.)

A solid 67% support for the development of an independent **European positioning system** was recorded, even though there was no Member State where a majority found this development "very important"; most such replies were given in Poland (44%), Hungary and Bulgaria (both 40%). Combining all affirmative replies – regardless of intensity – Polish (86%), Cypriot (82%), Hungarian and Finnish (79%) respondents were most likely to regard such developments as important to any extent.

Development of an independent European Positioning System (GPS)

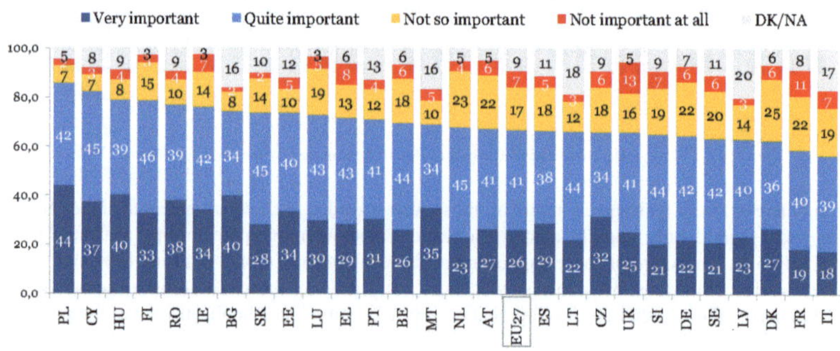

Q4. How important is in your view to develop the following space applications for Europe?
Base: all respondents, % by country

France (33%), Denmark (31%) and United Kingdom (29%) were the Members States where the highest share of public opinion considered the development of an independent European positioning system as being "not important", while in several Member States a significant minority could not formulate an opinion (e.g. 20% in Latvia, 18% in Lithuania).

From a socio-demographic perspective, this initiative was most solidly supported by men (73% vs. 61% women), the best educated (73%, in sharp contrast with those indicating the shortest schooling: 54%) and metropolitan residents (73%). The support levels decreased with age, but even those over 55 were significantly more likely than not to find the development of an independent positioning system quite or very important (62%, see Annex Table 5b.)

An overwhelming majority in most countries (most notably in Hungary, Romania, Ireland and Poland, about 80% in each) found the provision of **space-based monitoring for security applications**[4] important. However, significant minorities do not consider so important or important at all this type of application in a number of countries. Those who believe that space-based monitoring that improve citizens security are not so important, or not important at all for Europe reached 46 % in Austria, 43% in Germany, 39% in Sweden and 37% in Denmark and France.

[4] the questionnaire asked about such services in general, not specifying what they might be

Provide space-based monitoring tools to improve citizen security

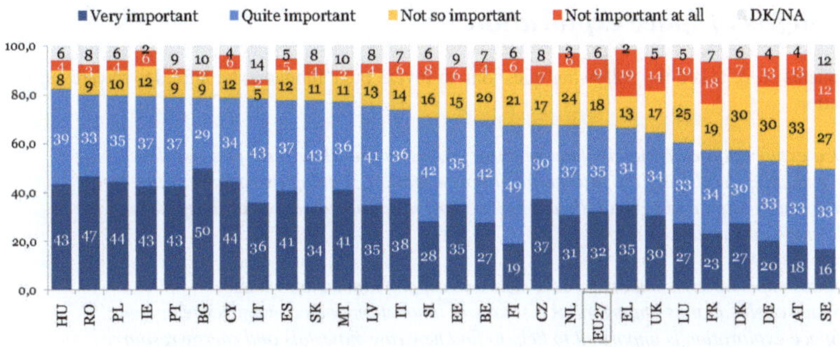

Q4. How important is in your view to develop the following space applications for Europe?
Base: all respondents, % by country

The provision of satellite services aimed at improved citizen security were regarded as least important (relatively speaking, as the majority in all segments still found such developments at least quite important) by those with degree-level education (63%, with 33% essentially rejecting the idea by stating that they find such developments unimportant). Metropolitan residents were also less likely than others to be attracted to such developments (important: 65%, not important: 31%). The highest approval of such developments was recorded among the youngest: 74% of under-25 year olds confirmed that these may be important developments, with 24% having had the opposite view. (See Annex, Table 8b.)

3. Space exploration

3.1 Benefits of space exploration

Respondents were asked the following questions to reveal their fundamental attitudes towards space exploration:

Space exploration activities include sending probes to the Moon and planets, developing space shuttle, developing the International Space Station, sending astronauts to the Moon, and travel to Mars and beyond. Please tell me how much do you agree or disagree with the following statements:

A. Space exploration is important as a means to expand human knowledge
B. Space exploration is important as a source of innovation and economic development
C. Space exploration is important to help to find new raw materials and energy resources
D. Space exploration is important to better protect our planet

While each aspect was regarded as generally important (the vast majority of respondents agreed with each), the ranking shown on the chart below indicated that citizens tend to link space research and exploration to more abstract benefits. 79% saw it as important due to its contribution to the expansion of human knowledge (about three in ten of those interviewed agreed strongly). 72% believed that space exploration may add to mankind's ability to protect the Earth, 71% believed that space exploration may lead to accessing new energy resources and raw materials, and 70% thought that space research can be a source of innovation and economic development.

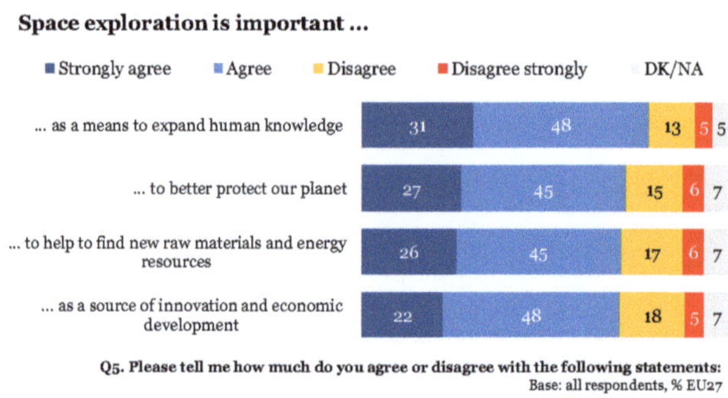

Less than a quarter of respondents disagreed that space exploration is important to achieve the stated objectives, as shown above.

The individual analysis of the various possible benefits of space exploration by Member State and by socio-demographic segments did not shed light on significant variations: all countries and all segments were in general convinced of the importance of the listed benefits, and broadly speaking the same countries and same segments believed most and least in the importance of each. Details are provided below.

The most widely acknowledged benefit of **space exploration** – that it **adds to human knowledge** – was most often cited by Bulgarian (93%), Hungarian (91%) and Greek (89%) respondents, while in Italy, France and the Netherlands, 'only' 72% of respondents agreed that this aspect was important.

Space exploration is important as a means to expand human knowledge

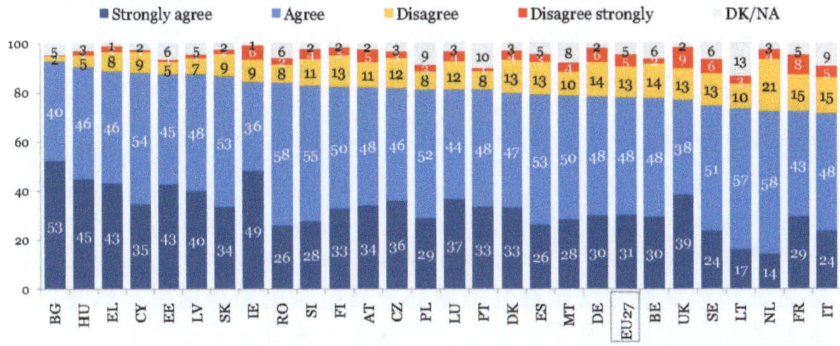

Q5. Please tell me how much do you agree or disagree with the following statements:
Base: all respondents, % by country

Men (81%), the younger generations (15-24 years: 85%, 25-39 years: 81%), those with degree-level education (84%) and metropolitan residents (83%) were more affirmative than the average (EU27: 78%) about the most abstract benefit of space exploration. (Annex, Table 9b.)

The technological advantage (that **space exploration is important because it facilitates innovation and thus economic development**) was once again most confirmed by those interviewed in Bulgaria (84%), Hungary (83%) while Ireland ranked third (79%). Least convinced of, though still largely in agreement with such a link were the French (60%), Swedish (64%), Italian and Maltese respondents (65% each).

Space exploration is important as a source of innovation and economic development

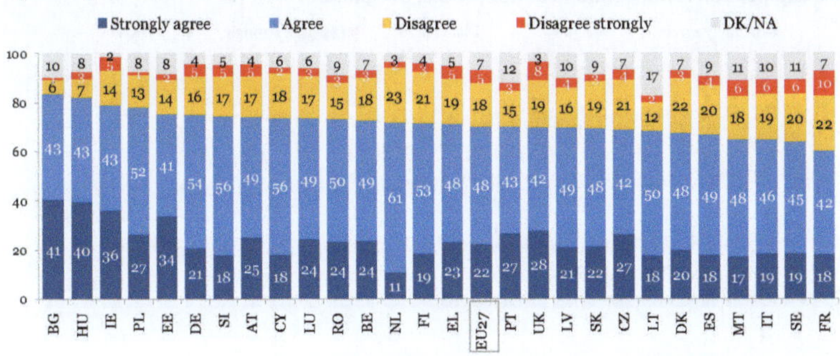

Q5. Please tell me how much do you agree or disagree with the following statements:
Base: all respondents, % by country

Compared to an EU average of 70%, it was once again men (75%), young people (15-24 years: 76%), those with degree-level education (75%) and metropolitan residents (74%) who appeared to see most clearly a link between space exploration and technological and economic benefits (See Annex, Table 10b.)

Bulgarians (85%), Romanians (84%), Poles and Cypriots (both 83%) most agreed that **space exploration is important in order to find new raw materials and energy resources**, while most scepticism in this regard was encountered in Austria (where only 56% agreed that space exploration is indeed important to achieve such outcomes), Sweden and France (both 63%).

Space exploration is important to help to find new raw materials and energy resources

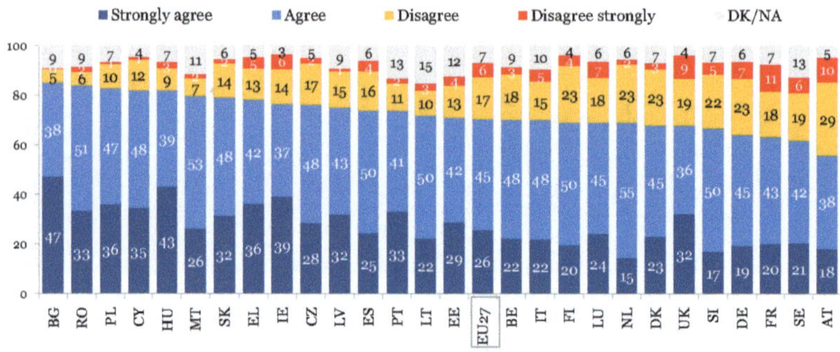

Q5. Please tell me how much do you agree or disagree with the following statements:
Base: all respondents, % by country

This aim of space exploration was most attractive (compared to a EU27 average of 71%) to the youngest respondents (15-24 years: 79%) and manual workers (77%). As shown in Table 11b in Annex, variation across segments remained rather limited (the level of agreement ranged between 64% and 75% in the rest of the segments).

Space exploration is important to protect planet Earth according to most people in Bulgaria once again (87%), followed by Poland (84%), Greece and Romania (83% both). On the bottom end of this ranking the survey found Austria (58%), The Netherlands (64%) and Sweden (65%).

Space exploration is important to better protect our planet

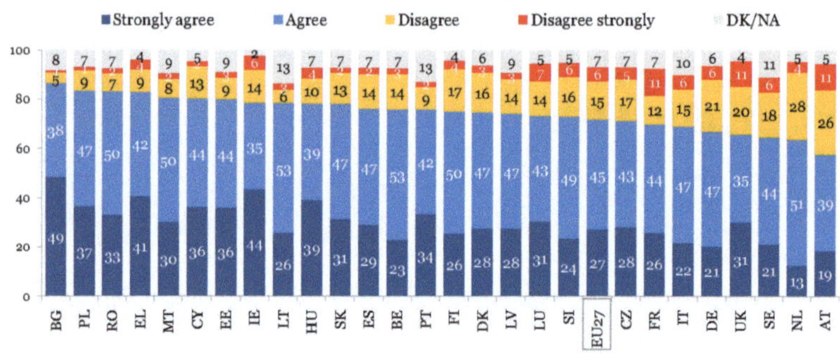

Q5. Please tell me how much do you agree or disagree with the following statements:
Base: all respondents, % by country

Only in the youngest segments was above-average agreement recorded (compared the 72% on EU27 level) as to the importance of space exploration in protecting our planet: 77% both of the 15-24 and 25-39 year olds acknowledged that this might be an important purpose of space exploration. (See Annex, Table 12b.)

3.2 Increased EU activity in space exploration

The question of whether **the European Union should do more in the field of space exploration** profoundly divides European citizens. 26% believe that the EU should *definitely* do more, while a similar number of respondents oppose further involvement (28%). A relative majority (38%) is not entirely convinced about the necessity, but feels that the EU should *perhaps* put more emphasis on this field. Overall, this signals that while a majority favours EU action in the field of space exploration, only some of the supporters seem to be whole-heartedly committed to this. 30% provided an outright negative response to this question (28% opposed more involvement and 2% indicated that it does not matter for them whether or not the EU is more active in space exploration).

Should the European Union do more in the field of Space Exploration?

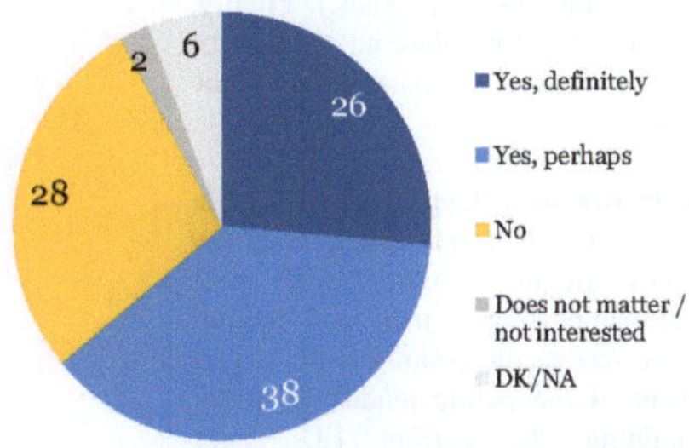

Q6. Should the European Union do more on the field of Space Exploration?
Base: all respondents, % EU27

Looking at the results in the various Member States, it appears that the highest levels of support for increased space exploration activities are found in countries that were most likely to acknowledge the importance of the various goals of space exploration: Bulgaria and Greece; in both countries more than half of those interviewed (55%) opted *definitely* for the EU being more active in this area. On the other hand, only about one in five respondents expressed full support in a number of various Member States, including Sweden (16%), France, Finland (both 19%), the Netherlands, Lithuania, Belgium (20% each) as well as in Italy, Slovakia and the Czech Republic (all 21%).

Should the European Union do more in the field of Space Exploration?

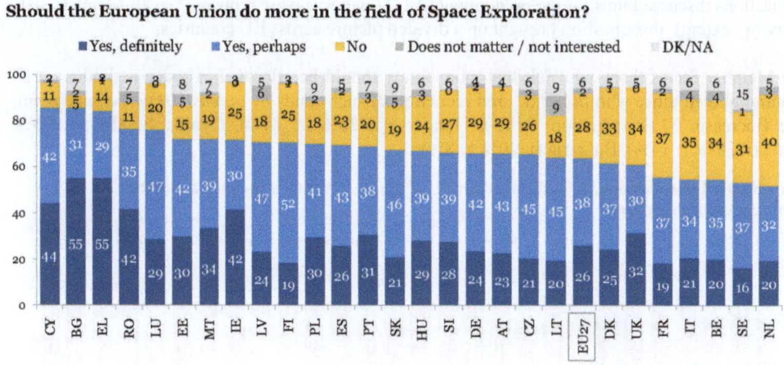

Q6. Should the European Union do more on the field of Space Exploration?
Base: all respondents, % by country

Nevertheless, the definite "no" answer was dominant only in the Netherlands (40%) and scored tied with "perhaps yes" in France (37%), Italy (35%) and Belgium (34%).

Considering the socio-demographic segments, whole-hearted support for increased EU activity in space exploration was well above average (that being 26%, considering the "yes, definitely" replies) among men (33% as opposed to 21% among women) and the youngest cohort (15-24 years of age: 35%). Slightly above average were the best educated and metropolitan Europeans (with 29% definite support in both groups, see Annex Table 13b.).

4. Budget priority

Overall there is strong support for EU funding of space activities at its current level (43%) or at increased level (20%). The 20% who felt that more budgetary resources should be allocated to space activities were outweighed – although only by a slim margin – by those who felt that the EU should reduce such spending (23%). (The survey was conducted under circumstances characterised by significant economic problems in essentially all Member States of the EU, resulting from the ongoing global economic and financial crisis.). Finally, 15% could not formulate an opinion or were not interested in answering the question.

The share of the European budget to cover all space activities including space exploration should be...

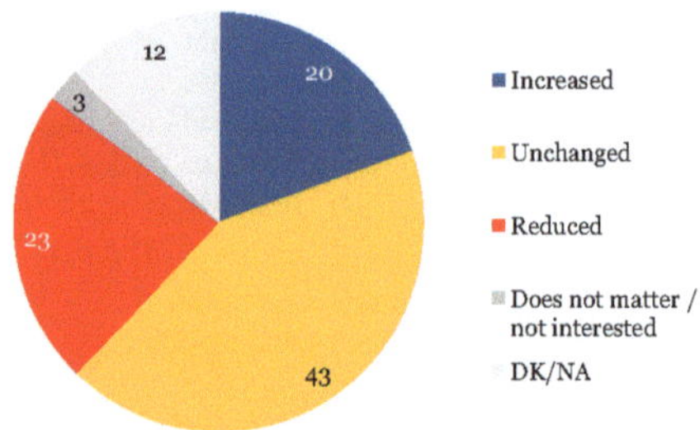

Q7. There are various budget priorities for the European Union. According to you, the share of the European budget to cover all space activities including space exploration should be:
Base: all respondents, % EU27

Only in Romania, Cyprus and Bulgaria was stability of the allocated funds *not* the most frequent answer (with most citizens supporting an increased budget) – in the rest of the countries the largest segment of the public tended to prefer maintaining the current EU space budget.

Reducing the European budget for space activities is an option shared by only 23% of the Europeans, with the highest scores recorded in France (32%), Spain (30%), Italy (27%) and United Kingdom (26%) and the lowest in Bulgaria (8%), Slovakia (12%) and Finland (13%).

While generally this study found only cosmetic variations of attitudes across Member States (that is, in most questions discussed thus far the predominant EU27 patterns were replicated in all Member States to a varying extent), this question brought up a divided picture across EU countries.

In five Member States those who felt that more funds should be allocated to space activities on EU level outnumbered those who rather desired reduction of such funds *by at least 10 percentage points*: Greece (increase: 31%, reduction: 21%[5]), Hungary (i: 26%, r: 14%), Cyprus (i: 38%, r: 16%), Romania (i: 37%, r: 14%), Bulgaria (i: 40%, r: 8%).

The share of the European budget to cover all space activities including space exploration should be...

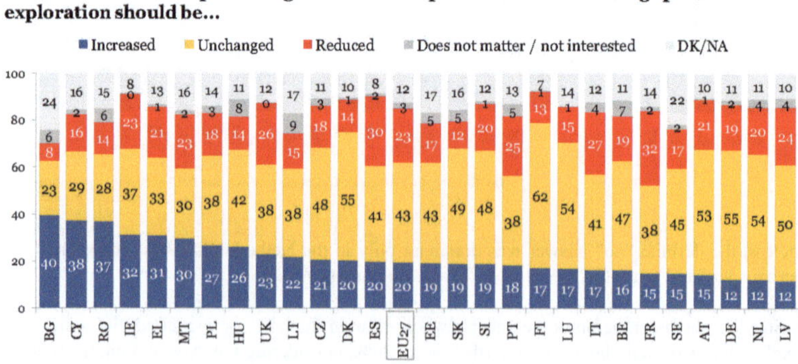

Q7. There are various budget priorities for the European Union. According to you, the share of the European budget to cover all space activities including space exploration should be:
Base: all respondents, % by country

[5] percentages for the other countries are provided respectively

In contrast, an opposite difference with a similar order of magnitude (10 percentage points) was also found, in four Member States: France (i: 15%, r: 32%), Latvia (i: 12%, r: 24%), Spain (i: 20%, r: 30%) and Italy (i: 17%, r: 27%).

Differences in preference manifest themselves in socio-demographic comparisons. While the plurality of respondents in each socio-demographic segment preferred that the EU space budget should be kept at its current level (see Annex, Table 14b.), the table below reveals different attitudes in those who prefer some change. For example men, the youngest respondents, and to a small extent the best educated were more of the opinion that funds should be increased, while women, elder respondents, those with only basic education and manual workers would prefer cuts in the EU space budget.

Space budget preferences by socio-demographic segments (Q7)

	% reduced	% unchanged	% increased
EU27	23	43	20
SEX			
• Male	20	41	25
• Female	25	44	14
AGE			
• 15 - 24	16	47	27
• 25 - 39	22	42	20
• 40 - 54	23	43	18
• 55 +	26	41	17
EDUCATION (termination age)			
• Until 15	33	36	14
• 16 - 20	24	44	18
• 20 +	18	43	21
URBANISATION			
• Metropolitan	22	41	23
• Urban	22	42	20
• Rural	24	44	17
OCCUPATION			
• Self-employed	22	37	24
• Employee	21	44	19
• Manual worker	28	42	19
• Not working	24	43	19
• Student	14	46	29

A.4 European identity through space: keywords for the future

Kai-Uwe Schrogl, Thomas Ballhausen, Blandina Baranes

A.4.1 European identity through space: transdisciplinary investigations

"Project Europe" can only succeed on the basis of a clear common European identity. European space activities are a building block for this existential mission by safeguarding our political, economic and cultural well-being.

Creating and shaping European identity through space is not the task only of engineers and scientists. It is a holistic endeavour comprising all academic and societal communities and encompassing all identified cognate disciplines in a transdisciplinary effort.

A.4.1.1 Three distinct "spaces can be identified"

European identity and global politics: How space technologies and applications are influenced by, and in turn create a consciousness of Europeanness.

European identity and the field of culture: How space, through the help of the humanities, creates other realities that disregard boundaries.

The future of European identity: How space can be used as an instrument for strengthening European identity and how, in turn, European identity may change under the influence of space exploration.

The potential of space for enhancing European identity is far from having been realised. Structuring and substantiating the benefits under these three "spaces" will contribute to their growing impact.

A.4.2 European identity through space: the conference

Under the auspices of the Swedish EU Presidency, ESPI organised the conference "European Identity Through Space" on 12–13 November 2009 in Vienna,

Austria. It was the first transdisciplinary approach to discussing space and its implications for shaping a European identity. At this event, European space policy experts met with representatives from the humanities for a wide-ranging dialogue at ESPI, where open-minded, creative discussions and joint efforts for a stronger Europe are encouraged and realised.

Participants in the conference included speakers from European academic institutions, politics, industry as well as international organisations. The results of the conference will be published in the series "Studies in Space Policy" edited by ESPI and published by SpringerWienNewYork. This flyer summarises the main results of the conference for use in policy debates.

A.4.3 First space

A.4.3.1 European identity and global politics

A.4.3.1.1 How space technologies and applications create an underlying consciousness of Europeanness

International context – The globalisation of European values and ideas can contribute to worldwide progress towards peace. The concept of Europe is rooted in but not restricted to the continent and its classical implications. The extension towards outer space and the major role of Europe in civilian space exploration inspire both cultural and technological achievements. Europe thus functions as a positive example and also a driver towards a global identity.

Technology – Ariane serves today as a prestigious symbol for numerous European cultural, scientific, technological and industrial achievements. Technological leadership and joint peaceful use of new possibilities support the implementation of a European identity that is aware of the benefits of space.

Applications – Europe's new and increasing role in international relations and as a leader in technological development is directly linked with major initiatives in space. Joint projects such as Galileo or GMES clearly demonstrate the significant role of independent European capabilities that, at the same time, stimulate global endeavours.

Security – Europe has changed from a continent of conflict to a union of peace. The view from space can bring a new awareness and tools for matters of internal and external security: Space has a security dimension and security has a space dimension. A positive and principled European space identity could play a formative role for a space related approach towards worldwide peace.

A.4.4 Second space

A.4.4.1 European identity and the field of culture

A.4.4.1.2 How space, through the help of the humanities, creates other realities which disregard boundaries

Philosophy – Europe is a desire and an idea – a passage from fear to hope. Europeanness, based on the philosophical concept of universalism, means transgressing frontiers and redefining relations between territories, limits and mobility. This never-ending intellectual curiosity, the process of identity formation and the longing for scientific progress, are reflected in its space ambitions.

The Arts – Europe is not only a leading pioneer of technological progress but also of passion and dreams. The arts, in all their diversity and historical dimensions, through their imaginative powers stimulate science and fiction well beyond conventional thinking. Examples such as the figurine "Cosmic dancer" encourage new ideas and communicate hard facts to a broader public in a unique way.

Education – Space works as a catalyst for diverse interests in science and research in many different educational programmes. Younger European generations can be reached and sensitised to engineering and science in general as well as to the social sciences through icons such as European astronauts.

Science communication – Communication about European space policy, space activities and related topics can be carried out via best practice networks, public events and educational programmes. These efforts should be optimised through European think thanks and specially trained Public Information Officers. They can provide a professional link between the scientific community and the public.

A.4.5 Third space

A.4.5.1 The future of European identity

A.4.5.1.3 How space can be used as an instrument for strengthening European identity

Dialogue – The European approach towards the future of joint space efforts needs to be transdisciplinary. Promoting space needs the involvement of new actors from the policy areas as well as from different scientific and cultural disciplines. The clear intention must be: distribution of European space achievements; further integration of space related topics in all relevant discussions; and the implementation or encouragement of vital forums for ensuring these dialogues on a European level.

Reflection – A better understanding of space will help to overcome boundaries. The European aerospace industry is a good example of the successful implementation of cross border cooperation and can stand as a role model. Corporate identity in space can lead to a corporate identity in Europe on a more general level.

Process – Europe is going through a permanent process of change: historically, geographically and geopolitically. It is an evolution which has to be revealed in a public, comparative and vivid discussion. The European space effort can act as a model case.

Visions – European space accomplishments are an anchor of European identity. Europe is a major actor in global politics, especially when it comes to space efforts. It is the best example of European coherence and a clear signal of its international competitive power. All areas of European life, ranging from politics to economics, from education to the arts, can benefit from our performance in space.

About the authors

Thomas Ballhausen is Key Researcher, LNO and Curator at the Filmarchiv Austria (Austrian Film Archive) and Associated Fellow at ESPI. He teaches Comparative Literature at the University of Vienna, his research centres on intermediality, film history and media theory. Ballhausen is, among other additional scientific activities, editor of the book series „exquisite corpse – Schriften zu Ästhetik, Intermedialität und Moderne" and „Sequenzen – Beiträge zu Bild, Sound & Text". He has published several scientific and literary books, among them „Kontext und Prozess. Einführung in eine medienübergreifende Quellenkunde (2005)", „Delirium und Ekstase. Die Aktualität des Monströsen" (2008), „Bewegungen des Schreckens" (2010) and „Bewegungsmelder" (2010).

Werner Balogh is the Programme Officer for Basic Space Technology in the United Nations Office for Outer Space Affairs which is based in Vienna, Austria. From 1997 to 1999 he gained first work experience with the United Nations as an Associate Expert for Space Applications. In this role he was involved in the planning and organisation of the Third United Nations Conference on the Peaceful Uses of Outer Space (UNISPACE III), held in July 1999. Prior to re-joining the United Nations in 2006 he worked as an International Relations Officer for the European Organisation for the Exploitation of Meteorological Satellites (EUMETSAT), representing EUMETSAT at the European Commission and as a Programme and Project Manager for the Austrian Space Agency. Werner Balogh holds engineering and doctorate degrees in technical physics from the Vienna University of Technology and master degrees from the International Space University and the Fletcher School of Law and Diplomacy.

Blandina Baranes joined the European Space Policy Institute (ESPI) in Vienna in February 2005 and currently holds the position of ESPI Communications Manager. Prior to this, she was the chief librarian of the Jewish Studies Department of the University of Vienna. During the past years she has also worked as a documentalist and librarian for different institutions, such as the Austrian Broadcasting Corporation, the *Der Spiegel* magazine and others. She conducted her studies and research in Austria and Israel and graduated with a Masters Degree from Vienna University, Faculty of Philosophy, Department of Social and Cultural Anthropology. She is co-editor of the past three volumes of ESPI's

"Yearbook on Space Policy", published several articles and participated as a speaker in a number of space related events on the topic of "Space and Interdisciplinarity".

Alan Belward works at the European Commission's Joint Research Centre in Italy where he is Head of the Global Environment Monitoring Unit in the Institute for Environment and Sustainability, a position he has held since 1998. The Unit provides information on terrestrial and marine ecosystem dynamics from satellite remote sensing to help in the sustainable management of natural resources, for assessing climate change impacts and for the conservation and use of biological diversity. He has a B.Sc. in Plant Biology from Newcastle University, UK and a Ph.D. in remote sensing studies of vegetation from Cranfield University, UK. From 2002 to 2006 he chaired the Global Climate Observing System's (GCOS) Terrestrial Panel. In 2009, he was invited to join the GCOS Steering Committee. He is currently a Principal Investigator with the NASA and USGS Landsat Data Continuity Mission Science Team and visiting lecturer to the Faculty of the Technical University of Vienna and the Diplomatic Academy of Vienna Postgraduate MSc Programme in Environmental Technology and International Affairs.

Ariane Cornell has served as the Executive Director of the Space Generation Advisory Council in support of the United Nations Programme on Space Applications (SGAC) since April 2009. She serves in the aerospace community in other capacities and organisations. Ariane Cornell is a member of Women in Aerospace's American and European branches. Further, she is a member of the American Institute of Aeronautics and Astronautics and serves on its International Activities Committee and the Young Professional Committee. Ariane also publishes and writes as a guest blogger for *Space News*. Prior to her current post, Ariane Cornell worked in management consulting, first with Accenture based in San Francisco and then with Booz Allen Hamilton in Washington, DC. With Accenture, she has lived and worked on IT projects in the Philippines, South Africa, Brazil, and the US. With Booz Allen Hamilton as a senior consultant in the aerospace and defense commercial consulting group, she helped develop strategies and solve operational issues for executives of the world's top aerospace and defense companies. Ariane Cornell is a French-American dual citizen whose interest in aerospace started at a young age: she earned her pilot's license at the age of 17 and spent many of her early internships at NASA Ames Research Center. She attended Stanford University where she earned her Bachelor of Science with Honors in Science, Technology, and Society with a focus in Management Science and Engineering. Her undergraduate honors thesis was entitled, "Sino-American Relations in Space: Cooperation, Competition, or Coopetition?"

Michael Gleason is the international relations division chief, Department of Political Science, United States Air Force Academy, Colorado. He teaches courses in U.S. National Space Policy, American Government, and International Relations. He is an Associate Director and founding team member of the Eisenhower Center for Space and Defense Studies.
As a former spacecraft mission controller he has extensive experience in the U.S. Defense Meteorological Satellite Program, the Defense Satellite Communications System, the Defense Support Program, and the Milstar satellite system.
Michael Gleason began teaching political science at the United States Air Force Academy in 2002. He earned his Ph.D. at George Washington University, Washington D.C. in 2009. His Ph.D. dissertation examined the European Galileo satellite program and the European space sector. Michael Gleason has many publications to his credit in high ranking journals and books.

Peter Habison former Director of the Vienna Planetarium, Kuffner- and Urania Observatories in Vienna, Austria. He studied physics at the University of Technology in Vienna, astronomy and the history of sciences at the University of Vienna (PhD). Moreover he also holds a degree in "Management and Education". He gathered and work experience at University of Innsbruck, Université Libre at Brussels (Belgium) and the Instituto Astrofisica de Canarias at Tenerife (Spain). He has experience at the University of Technology in Vienna, as a tutor at the FFG Astrophysical Summer School Alpbach and he gives regular lectures at planetaria, observatories, universities and conferences worldwide on astronomical topics. He is an expert in the history of Kuffner Observatory and Kuffner family, Planetarium techniques, astronomy visualisation, communicating astronomy with the public and science communications. He is active in stimulating interest and enthusiasm for the young generation in space astrophysics and natural sciences. He is the author of more than 40 publications and he is member of the Österreichische Physikalische Gesellschaft, Astronomische Gesellschaft and International Academy of Astronautics, Rat Deutscher Planetarien, Gesellschaft Österreichischer Planetarien and since 2009 the Austrian delegate to the ESO Science Outreach Network.

Marcus Hornung is lecturer at the Institute for European Studies of the Chemnitz University of Technology, focussing on European space policy. He holds a master's degree in European Studies of the Free, the Humboldt and the University of Technology Berlin. In 2010, Marcus Hornung has been working for the European Space Policy Institute as an intern. He furthermore holds a bachelor's degree in European Studies from the Chemnitz University of Technology, including studies in Chemnitz and Brno, internships in Vienna and

Berlin and assistant work for European Integration and Social and Economic Geography chairs. Besides his studies he has been working for the European Academy Berlin since 2009, organising and assisting seminars, field trips and workshops on European politics.

Ulrike Landfester is at present (since 2006) Member of the Standing Committee for the Humanities at the European Science Foundation, since 2004 Member of the Research Council of the Swiss National Foundation (Humanities and Social Sciences) and since 2003 full professor for German language and literature at the University of St. Gallen (Switzerland). She has obtained a Ph.D at the University of Munich (Dissertation: Der Dichtung Schleier. Zur poetischen Funktion von Kleidung in Goethes Frühwerk – The Veil of Truth. The poetical function of clothing in Goethes early work), undertaken the study of German literature, English literature and Medieval literature at Albert-Ludwigs-Universität in Freiburg (Germany). Prior to that Ulrike has undertaken the study of Archeology, Egypteology and Early History at the aforementioned university in Freiburg.
She has been the co-editor of the complete works of Rahel Levin Varnhagen (Edition Rahel Levin Varnhagen), University of Hamburg. Ulrike has also taught at several institutions; inter alia, University of Vienna (Austria), professor at the University of Konstanz, full professor at the University of Frankfurt am Main.

Stephan Lingner is Vice Director of the „Europäische Akademie zur Erforschung von Folgen wissenschaftlich-technischer Entwicklungen Bad Neuenahr-Ahrweiler GmbH“ since 2005. He is senior scientist with specific responsibility for the academy's research programme on technology and environmental assessment and has coordinated several interdisciplinary projects in this field. He is also the Managing Editor of the Springer journal "Poiesis & Praxis: International Journal of Ethics of Science and Technology Assessment". Previously, he was a research fellow at the German Aerospace Center (DLR) at Cologne, where he performed systems and strategic studies on new options for spaceflight and space exploration. Before, he was planetary scientist at Münster University, while analysing lunar rock samples from the Apollo 14 landing site. Stephan Lingner has published numerous scientific articles and reports on space science and exploration as well as on environmental protection and technology assessment. He holds a doctoral degree in planetary chemistry and worked as an Expert Reviewer for the Intergovernmental Panel of Climate Change (IPCC) and on other scientific programmes, projects and papers. Stephan Lingner was lecturer for Ecology at the Koblenz University of Applied Sciences and has been member of the German speaking

„Netzwerk Technikfolgenabschätzung" (NTA) and of the European Space Policy Research and Academic Network (ESPRAN).

Giulia Pastorella is currently studying a European Affairs Double Master's degree offered by the Institut d'Etudes Politiques (Paris) and the London School of Economics (London). She holds a bachelor's degree from the University of Oxford in Philosophy and Modern Languages. She has been a Research Assistant at the European Space Policy Institute in August–September 2010, where she co-authored an issue in the Occasional Paper series "ESPI Perspectives". She is currently an editor for Barometer Intelligence Ltd, and she previously did an internship as a research and archive assistant at Le Monde. She studies at the London School of Economics and Political Science while also working as an Academic Assistant at Sciences Po in Paris.

Wolfgang Rathgeber from Germany studied Electrical Engineering at the Universities of Erlangen (Germany), Campinas (Brazil) and Valencia (Spain). He has been with the German Aerospace Centre (DLR) since 1999. Until 2003, he was a doctorate student at the Institute of Radio Frequency Engineering and Radar Systems in Oberpfaffenhofen (Germany) with a thesis on high resolution signal processing for Synthetic Aperture Radar (SAR). From 2003 until 2005, he was scientific staff member at the European Academy of Technology Assessment Bad Neuenahr-Ahrweiler (Germany). From 2005 to 2010 he was seconded to the European Space Policy Institute (ESPI) in Vienna (Austria). From 2010 on, he has been detached to the European Space Agency (ESA) in Paris.

Nina-Louisa Remuss was Resident Fellow at the European Space Policy Institute (ESPI) and is currently with DLR German Aerospace Center, Berlin. In addition she works as a Research Assistant for a German MP at the German Bundestag since July 2010. She has been contributing to ESPI's Research Programme Space and Security since July 2008. In particular, she co-authored a study on Europe's role in the peaceful-uses of outer space debate, led a study and a related workshop on the contribution of space applications to internal (i.e. homeland) security which was conducted under the auspices of the Czech EU Council Presidency, and further led two studies on Responsive Space and the contribution of space applications to the fight against piracy. She has contributed numerous articles and papers to leading journals in the field and is regularly invited to speak at conferences in Europe and the U.S. In the context of her research she has been organising workshops and conferences with the participation of leading personalities in the respective fields (from European institutions, agencies and industry) where she also has been acting as moderator or session chair. In 2009 she was invited to become member of the Panel of Experts of the EU Framework Programme 7 project STRAW. (Security Technology Active Watch), which provides a building block towards the development of a European Security Network (ESN). Also in 2009, she was tasked with the conduct of ESPI's support of the Presidency of the European Interparliamentary Space Conference (EISC). She holds a Bachelor Degree in European Studies from the University of Maastricht (The Netherlands), a Master's Degree in International Security Studies from the University of St. Andrews (United Kingdom) and spent an exchange semester at the University of Bologna (Italy) and the Nilsson Center for European Studies of the Dickinson College (Italy).

Kai-Uwe Schrogl former Director of the European Space Policy Institute (ESPI) in Vienna, Austria is currently Head of Policies Department at ESA, Paris. Prior to this, he was the Head of the Corporate Development and External Relations Department in the German Aerospace Center (DLR). Previously he also worked with the German Ministry for Post and Telecommunications and the German Space Agency (DARA). He has been a delegate to numerous international forums and recently served as the chairman of various European and global committees (ESA International Relations Committee and two UNCOPUOS plenary working groups). He presented, respectively testified, at hearings of the European Parliament and the U.S. House of Representatives. Kai-Uwe Schrogl has written or co-edited 12 books and more than 130 articles, reports and papers in the fields of space policy and law as well as telecommunications policy. He is editor in chief of the "Yearbook on Space Policy" and the book series "Studies in Space Policy" both published by ESPI at SpringerWienNewYork. In addition he sits on editorial boards of various international journals in the field of space policy and law (Acta Astronautica, Space Policy, Zeitschrift für Luft- und Weltraumrecht, Studies in Space Law/Nijhoff). Kai-Uwe Schrogl is a Member of the Board of Directors of the International Institute of Space Law, Member of the International Academy of Astronautics (recently chairing its Commission on policy, economics and regulations) and the Russian Academy for Cosmonautics. He holds a doctorate degree in political science and lectures international relations at Tübingen University, Germany (as an Honorary Professor). He has been a regular lecturer at, for example, the International Space University (where he serves as Adjunct Faculty) and the European Centre for Space Law's Summer Courses.

Mathias Spude is the Astrium Director Communication & Public Relations Germany since 1999, based in Ottobrunn and Bremen. As of 1994 he was Deputy Managing Director of the German Aerospace Industries Association BDLI in Bonn in charge of Space, Aeronautics and Defence. Before, he was Head of ESA Controlling at the then German Space Agency DARA in Bonn. In this function he represented Germany in several ESA Committees. In 1989 he acted as the first ever secretary of the European Centre for Space Law at ESA in Paris. He has published numerous articles and documents on space activities, participated to the „Handbuch des Weltraumrechts" and elaborated on space activities as governmental task. He is member of the International Academy of Astronautics, the International Institute of Space Law, the European Centre for Space Law, and the German Society of Aerospace. He holds a legal doctorate degree at the University of Cologne and graduated at Freiburg University.

Gerhard Thiele is Resident Fellow at the European Space Policy Institute (ESPI) in Vienna, Austria, since April 2010. His research focuses on Human Space Exploration and questions related to European Autonomy in Space. Before joining ESPI he worked as astronaut first for the German Aerospace Center (DLR), then as of 1998 for the European Space Agency (ESA). In February 2001 Gerhard Thiele participated in the Shuttle Radar Topography Mission (SRTM) onboard space shuttle Endeavour, which generated a digital three-dimensional model of the Earth's landmass. From 2005 until 2010 he served as Head of the Astronaut Division of the ESA in Cologne, Germany, where he was also responsible for the selection of the last ESA astronaut class in 2009. Gerhard Thiele has co-edited books on Man in Space and published articles, reports and papers on human exploration as well as occupational psychology regarding human factors in space flight. Gerhard Thiele is a member of several professional organisations and supported the Lunar Development Forum of the International Astronautical Federation (IAF). He holds a doctorate degree in physics and has been a guest lecturer i.a. at Umea University, Sweden, and Universität Karlsruhe, Germany.

Ignatius van Neerven is a philosopher and theologian. He studied in the Netherlands and followed courses in Germany, Belgium and the United States. He lectured on philosophy of technology, on philosophy of culture, on business ethics and spirituality in business at Tilburg University and at the Polytechnic Institute of the Hague/Rijswijk in the Netherlands. He gave lectures and organised conferences in the Netherlands, Germany, France, Poland and Belgium. He published translations of philosophical and theological works. During the last ten years he worked as deacon in the parish of Bergen op Zoom, Netherlands.

Nowadays he lives in St. Cloud, France, where he continues his research and teaching activities.

Christophe Venet is a Ph.D. candidate at Sciences Po, Paris, and a Research Associate at the Institut Français des Relations Internationales (Ifri) in Paris since October 2010. In 2009 and 2010, he worked at the European Space Policy Institute (ESPI), first as a Research Assistant and then as an Associate Fellow. In this capacity, he contributed to the Report "Space Policies, Issues and Trends in 2008/2009" and collaborated as a co-editor and co-author to the "Yearbook on Space Policy 2008/2009". He also worked on the issue of space commerce and space entrepreneurship, preparing several presentations and articles on these issues. He was invited to become a peer-reviewer for the journal "Acta Astronautica" in August 2009. He graduated from the Institut d'Etudes Politiques (IEP) de Strasbourg, France, and studied international relations at the Moscow State Institute of International Relations (MGIMO), Russia. He also holds a Masters degree in Peace Studies and International Politics from Tübingen University. His dissertation deals with EU policy in the field of space and security, focusing on the actorness of Europe and on interests and norms underlying the policy processes within the European Space Policy.

Moniel Verhoeven studied the problems of the migration from West-Africa to Europe, doing field research in Mali, Mauretania and Senegal as well as in Paris and its outskirts. Socio-economical and (mental) health problems, expressed in symbolic communication, were her key points of interest. Last half year she worked for the SAMU-social in Paris, an organisation for aid and shelter to the homeless and refugees. She reported to the French Department of Employment and published several articles in France and in the Netherlands.
As Marketing Researcher for SKF-France, she was responsible for the Benelux. She investigated new industrial, technological and economical developments. This generated her interest for space. At the Aerospace Department of the University of Technology in Delft, the Netherlands, she researched the problems of intercultural management in the aerospace industry. She created the Round Table Conferences Jean Monnet, in cooperation with the ESA, The Delft University and the Association Jean Monnet.
At this moment she is employed in Paris in the after-sales sector for Air-France/KLM and is teaching International Marketing at the business school Negocia/Advancia.
She teaches also intercultural communication and rituals in business, education and health care. She lectured in the Netherlands (Nijmegen, Eindhoven, Breda),

in France (Montpellier, Paris) and in Germany (Darmstadt). On this subject she published several articles in Great Britain, France and the Netherlands and the handbook for the Dutch National Healthcare.
At the Radboud University of Nijmegen, the Netherlands, she studied Anthropology, Philosophy and Cinematography. In France she followed the post-MBA program *Mastering Change: Developing coaching and consulting skills* at the INSEAD in Fontainebleau.

Speakers of the conference from left: Kai-Uwe Schrogl, Mathias Spude, Sally Jane Norman, Jacques Blamont, Alan Belward, Wolfgang Rathgeber, Blandina Baranes, Stephan Lingner, Moniel Verhoeven, Johannes von Thadden, Nina-Louisa Remuss, Jörg Kreisel, Werner Balogh, and Peter Habison.

About the authors

List of acronyms

A

AMNH: American Museum of Natural History
AD: Anno Domini
APRSAF: Asia-Pacific Regional Space Agency Forum
APSCO: Asia-Pacific Space Cooperation Organisation
ASAT: Anti Satellite Weapon
ATV: Automated Transfer Vehicle

B

BC: Before Christ
BIRD: Bispectral Infra Red Detection

C

CD: Conference on Disarmament
CEOS: Committee on Earth Observation Satellites
CERN: European Organisation for Nuclear Research
CNES: Centre National d'Etudes Spatiales, French Space Agency
CoC: Code of Conduct
COP: Common Operational Picture
COPUOS: Committee on the Peaceful Uses of Outer Space
CSDP: Common Security and Defence Policy

D

DASA: Deutsche Aerospace Aktiengesellschaft, former German Space Agency
DG: Directorate General
DLR: Deutsches Zentrum für Luft- und Raumfahrt, German Aerospace Center
DMC: Disaster Monitoring Constellation
DU: Digital Universe

E

EAC: European Astronaut Centre
EADS: European Aeronautic Defence and Space Company
EB: Eurobarometer
EC: European Commisison
ECV: Essential Climate Variable

EDA: European Defence Agency
EDRS: European Data Relay Satellite
EEAS: European External Action Service
EEG: Eastern European Group
EGNOS: European Geostationary Navigation Overlay Service
ELDO: European Launcher Development Organisation
ENP: European Neighbourhood Policy
ENVISAT: Environmental Satellite
EO: Earth Observation
EP: European Parliament
ePOD: ESO education and Public Outreach Department
ERNO: Entwicklungsring Nord
ERS: European Remote-Sensing Satellite
ESA: European Space Agency
ESDP: European Security and Defence Policy
ESERO: European Space Education Resource Offices
ESO: European Southern Observatory. European Organisation for Astronomical Research in the Southern Hemisphere
ESON: ESO Science Outreach Network
ESP: European Space Policy
ESPI: European Space Policy Institute
ESRAB: European Security Research Advisory Board
ESRIF: European Security Research Innovation Forum
ESRO: European Space Research Organisation
ESRP: European Security Research Programme
ESS: European Security Strategy
EU: European Union
EUFOR: European Union Force
EUMETSAT: European Organisation for the Exploitation of Meteorological Satellites
EURATOM: European Atomic Energy Commission
EURISY: European Association for the International Space Year
EUSC: European Union Satellite Centre
EUTELSAT: European Telecommunications Satellite Organisation

F

FAO: Food and Agriculture Organisation
FAZ: Frankfurter Allgemeine Zeitung
FMCT: Fissile Material Control Treaty
FTSE: Financial Times Stock Exchange

G

G7: Group of Seven
GCOS: Global Climate Observing System
GIANUS: Global Integrated Architecture for Innovative Utilisation of Space for Security
GIS: Geographic Information System
GMES: Global Monitoring for Environment and Security
GNSS: Global Navigation Satellite System
GOCE: Gravity Field and Steady-state Ocean Circulation Explorer
GoP: Group of Personalities
GPS: Global Positioning System

I

IAA: International Academy of Astronautics
IAC: International Astronautical Congress
IADC: Inter-Agency Space Debris Coordination Committee
IAEA: International Atomic Energy Agency
ICG: International Committee on Global Navigation Satellite Systems
IGMASS: International Global Aerospace Monitoring System
IRC: International Relations Committee
ISRO: Indian Space Research Organisation
ISS: International Space Station
IYA: International Year of Astronomy

J

JRC: Joint Research Centre of the European Commission

M

MBB: Messerschmitt-Bölkow-Blohm
MUSIS: Multinational Satellite-based Imagery System for Surveillance, Reconnaissance and Observation

N

NASA: National Aeronautics and Space Administration
NEO: Near-Earth Objects
NEREUS: Network of European Regions Using Space Technologies
NGO: Non Governmental Organisation
NPS: Nuclear Power Sources
NRC: National Research Council

O

OHQ: Operations Headquarters

P

PAROS: Prevention of an Arms Race in Outer Space
PASR: Preparatory Action for Security Research
PECS: Plan for European Cooperating States
PNR: Passenger Name Record
PPWT: Treaty on the Prevention of the Placement of Weapons in Outer Space, the Threat or Use of Force against Outer Space Objects
PR: Public Relations

R

R&D: Research and Development

S

S&T: Science and Technology
SatCom: Satellite Communication
SatNav: Satellite Navigation
SPASEC: Panel of Experts on Space and Security
SPOT: Satellite Pour l'Observation de la Terre, Earth Observation Satellite
SSA: Space Situational Awareness
START: Strategic Arms Reduction Treaty
STEM: Science, Technology, Engineering and Mathematics
STM: Space Traffic Management

T

TEU: Treaty on European Union
TV: Television

U

US: United States
UK: United Kingdom
UNESCO: United Nations Educational, Scientific and Cultural Organisation
UNFCCC: United Nations Framework Convention on Climate Change
UNIDROIT: International Institute for the Unification of Private Law
UNISPACE: United Nations Conference on the Exploration and Peaceful Uses of Outer Space
UNOOSA: United Nations Office for Outer Space Affairs

UN-SPIDER: United Nations Platform for Space-based Information for Disaster Management and Emergency Response
UPI: United Press International

W

WEOG: Western European and Others Group
WW: World War

List of figures and tables

Figures

Chapter 1 Perspectives from space studies

Chapter 2 Perspectives from the humanities

Tables

Chapter 1 Perspectives from space studies

MIX
Papier aus verantwortungsvollen Quellen
Paper from responsible sources
FSC® C105338

If you have any concerns about our products,
you can contact us on
ProductSafety@springernature.com

In case Publisher is established outside the EU,
the EU authorized representative is:
Springer Nature Customer Service Center GmbH
Europaplatz 3, 69115 Heidelberg, Germany

Printed by Libri Plureos GmbH
in Hamburg, Germany